JN111579

Round-Table Talks of Whole Silks

絹大好き3

Love silk 3

挚爱丝绸3

織る・編む・着る&食べる

Weave, Knit, Wear & Eat

编·织·衣与食

中山れいこ／著　赤井弘／総監修　長島孝行／監修

Written and edited/NAKAYAMA Reiko　General supervised/AKAI Hiromu　Supervised/NAGASHIMA Takayuki

中山 令子／作者　赤井 弘／综合监督　长岛孝行／监督

日・英・中、3か国語表記 Written in Japanese, English and Chinese

日・英・中，三语言撰写

＊写真は、太繊度低張力繰糸の糸（P.90）/ Picture of large size less stress reeling yarn(P.90) / 图为低张力缫丝粗纤度纱(第90页)

繭と織物の間 / Process from cocoon to cloth / 茧与织品之间

どうしたら織物ができるの？/ How can cloth be produced？/ 如何制成织品？

カイコを飼育して最後に「繭」ができると、「この固い「繭」が本当に糸や織物になるのか?」と考える人や、教室で飼育した50～100個の「繭」で、糸や織物は作れないと考える人も多いでしょう。

この本では、小学校５年生の「私の飼育したエリサン*の繭から糸をとって何かを織りたい」との願いをかなえるため、簡単な道具で大切な少しの糸を織る織り方を紹介しました。

織ることの基本は、経糸と緯糸の組み合わせです。下の図のように、厚紙に糸をクルクル巻きつけてください。これが経糸です。

When silkworms are bred and finally they make cocoons, some breeders doubt the possibility to make thread and then cloth from this hard "cocoon, and some think of the possibility of making thread and then cloth from 50-100 cocoons bred in the classroom.
In this book, we introduced how to weave a small volume of thread out of a small amount of precious yarn with a simple tool, in order to fulfill the wish of 5th grade elementary pupils of weaving something from the yarn taken from Eri silkworm* cocoons bred by she.

The basis of weaving is a combination of warp and weft. Roll the thread around the thick paper as shown in the left figure. This is warp.

看着在教室中养殖的50～100只蚕宝宝变成蚕茧，或许很多人都会有这样的疑问:“真的有可能用这种坚硬的【蚕茧】来制造线和织物吗？”

本书主要介绍了如何使用简单的工具，编织珍贵的蚕丝，满足五年级小学生们:“我想从自己养的Eri silkworm(樗蚕)*制成的茧中抽出一条线，织出一些东西。”的愿望。

编织的基础是经纱和纬纱的组合。如左图所示将线绕在厚纸上，这就是经纱。

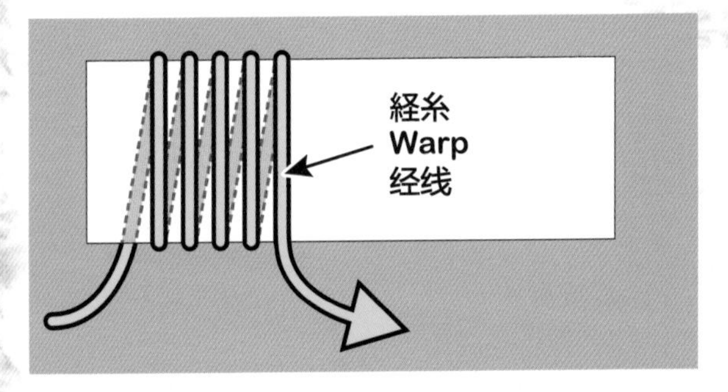

次に、この経糸を１本おきにすくって、糸を通してみましょう。これが緯糸です。

Then let's pick up warps every second lines and pass threads. This is the weft.

然后，每隔一次舀一次，让线程通过。这是纬纱线。

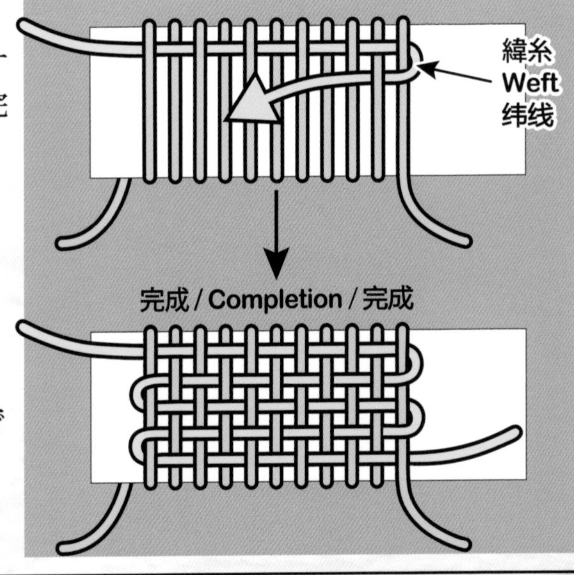

緯糸を端まで通すと、小さな織物が完成します。

小さすぎますか？

このように初心者が取り組みやすい、身近な材料や道具で簡単な織機を作り、帽子やバッグも紹介しました。

When you pass threads to the edge, a small cloth is completed.

只要将线穿到底，一个小小的织品就完成了！

Is it too small？

太小了吗？

Like this, we introduced an easy method for the beginners to make a simple loom with familiar materials and tools near at hand and then make hats and bags.

就是正如，本书主要是用身边熟悉的材料，或者工具，给大家介绍初学者也能简单上手的编制方法，比如帽子、包包的织品。

*エリサン：野蚕、『絹大好き２まゆの秘密』P.71参照。

*Eri silkworm: Wild silkworm, "Love Silk 2 Miracle of Cocoon", see page 71.

*樗蚕: 野生蚕，" 挚爱丝绸２茧的秘密 "，请参阅第71页。

シルクの手編み糸を編む / Knitting hand-knitted silk yarn / 我针织手编丝线

「手編み糸」と聞くと、多くの人はスキーやスケート用のざっくりとしたセーターを思い浮かべるでしょう。虚弱体質の筆者は、2013年ころから健康のために、絹糸を編んで着るようになりました。

しかし手編み用の絹糸は少なく、完成した見本や、絹ニットの手入れ方法を示す資料はほとんどありません。

市販にはない絹のバルキーセーターは、ウールのような太糸ではなく、織物用の細糸の引き揃えで編みました*。

この本では過去7年間で出会った、糸の扱い方などの困りごとや解決法を紹介しました。

When you hear "hand knitted yarn", many people may think of a rough and bulky sweater for skiing and skating. The present writer with fragile health has got to knit silk yarn by myself for my health since around 2013. There are few silk yarns for hand knitting, and there are few completed samples and references showing how to take care of silk knit.

Silk bulky sweater that is not available on the market is knitted not with a thick yarns like wool, but with thin ones after their alignment*. In this book, we have introduced problems and solutions on such as how to handle threads that we have encountered in the past 7 years.

当我们听到“ 手工编织纱线 ”,很多人第一时间想到的可能都是滑雪，滑冰时候用的粗糙的毛衣。体质虚弱的作者，从 2013 年左右，为了健康一直都穿自己手工编的丝绸毛衣和衣物。而手工编织用的蚕丝就少了，无论是已经完成的样品，还是材料入手的介绍，几乎都任何現存的资料没有。市面上所没有的丝绸膨体纱线毛衣，不是像羊毛那样用粗线织成的，而是用对齐许多细蚕丝线针织＊。在这本书中，我将向大家介绍，这七年来我在蚕丝针编织的这方面遇到的问题以及对应的解决办法。

「1年中絹を着て暮らす」お手入れ法
“Living with silk fabric all the year round”, Care for fabric / 丝绸衣物的日常保养方法

人の肌に近い優しい絹の繊維について、はるか昔の物語から、健康シルクの利活用まで、歴史を織り混ぜながら丁寧に書きました。また、気楽にゆったりと、シルクライフを楽しめるように、洗濯やアイロン掛け、毎日のお手入れ方法を、専門家に取材しました。

About gentle silk fiber which is close to human skin, I wrote carefully an article from the distant past matter to the recent usage for health care, including historic events. I interviewed experts about washing, ironing and daily care of silk products so that you can enjoy your silk life comfortably and at ease.

贴近人体皮肤的，我在编织中关于温柔的蚕丝纤维，并結合历史的演变我亡人真地写了比文章，从很久以前的故事到直到利用健康的蚕丝。我们就有关丝绸的洗涤，熨烫，和日常护理采访了有关专家，以便您可以享受丝绸生活的乐趣。

食べる・塗る / Eating and putting on / 食·美丽

シルクプロテイン入りスイーツや料理はどのような味か、また、シルクプロテインを飲んだ体験談など、日々の生活にプラスになるように、身近な内容にしました。

絹関連の日用品を製造する事業者は、日本全国でも珍しく、生産されている商品がデパートなどで見られるのはごく一部です。この本の巻末に、全国の皆様から広告を頂き、製品紹介もしました。ぜひご参考にしてください。

多くの研究者、絹関連産業の関係者にご協力を頂き、読者のお役に立つことを願い、この本を書きました。

We picked up the familiar topics such as how the silk protein containing sweets and/or dishes taste, the experience of drinking silk protein, and so on, in order that they may be useful for the daily life. The company that manufactures silk related daily commodities can be rarely found in Japan and commercial products can be found only at some limited departments. At the end of this book, products are introduced based on the advertisements from all over Japan. Please use them as the references. I wrote this book with the cooperation of many researchers and people involved in the silk related industry in the hope of helping readers.

我们将内容设为熟悉的，以便对我们的日常生活产生积极影响，例如丝绸蛋白的甜食和菜肴的味道以及服用丝绸蛋白的经验。在日本，与丝绸相关的日用品制造商少之又少，在百货商店中，只出售一些产品。在本书的最后，我们收到了来自全国各地的广告，并介绍了我们的产品，请参考。我和在丝绸公司及许多丝绸行业的研究人员的合作下，写了这本书，希望能对您有所帮助。

2021年7月 中山 れいこ / July 2021 NAKAYAMA Reiko / 2021年7月 中山 令子

＊引き揃え糸：太さ1㎜くらいの糸を3～5本使い、引き揃えて編みます。

＊Aligned thread: Align 3 to 5 threads of each about 1 mm thickness and knit them.

＊对齐许多细线：使用3到5根厚度约1毫米的线对齐编织。

もくじ / Table of contents / 目录

Chapter 1 Making Cloth / 第一章 做布

織ると編む / Weaving and knitting / 编与织

編物の始まり / Beginning of knit / 编品的起源

人の身体には羽や毛皮がないので、身体を包む布が必要です。200万年前の旧石器時代から、人々は細いひも状のものを絡み合わせたり結んだりして編み、身の回り品を作ってきたようです。

日本では、「福井県の鳥浜貝塚」の集落遺跡(約12,000〜5,000年前)から、大麻や縄、漁網、かごなどの植物繊維が出土しています。

Since the human body has no feathers nor fur, something to cover it is needed. Since the Paleolithic age of about two million years ago, people seem to have been making something to protect themselves by interknitting and/ or binding thin string-like material together. In Japan, plant fibers such as hemp or rope, fishing net and baskets have been excavated from the "Torihama Shell Mounds within ruins of a village in Fukui Prefecture" (about 12,000 to 5,000 years ago).

由于人类身体没有羽毛或皮毛，因此需要用布来包裹身体。从200万年前的旧石器时代开始，人们似乎通过缠绕和绑细绳子一起来制作日常用品。日本在【福井县鸟滨贝壳丘】的遗址(约12000～5000年前)中，出土了麻，绳，渔网，篮子等的植物纤维。

織物の始まり / Beginning of woven fabrice / 织品的起源

紀元前5000年ころのエジプトの出土品には糸を紡(つむ)ぐ道具があります。人々は布を織って、着ていたのでしょう。

また、中国では紀元前3500年頃には絹織物が生産され、紀元前300年ころの出土品に経錦(写真右)があります。

日本に養蚕が伝わったのは紀元前200年ころと考えられていますが、織機の出土は3世紀以後の弥生時代からです。

経錦(たてにしき)* / Vertical brocade / 垂直锦缎

中国戦国時代の衣服用絹布断片(個人蔵)
A silk cloth fragment during China Warring States Period (private collection)
中国战国时期，丝织品衣服的片段(私人收藏)

Among Egyptian artifacts around 5000 BC, there are tools for spinning threads. People would have woven and wore cloth. In China, silk fabrics were produced around 3500 BC, and one of the excavated items around 300 BC is " Vertical brocade (left photo)".

Sericulture might be transmitted around 200 BC, but discovered loom is one after 3rd century (Yayoi period).

在公元前5000年左右的埃及出土文物的发掘中，有用于纺纱的工具，大概当時人们会用编织布来穿着。中国在公元前3500年左右，已经有丝绸织品了。左图的垂直锦缎，就是在公元前300年出土的文物。

据信，公元前200年左右，养蚕业开始进入日本，但发掘的织机产品，是从三世纪以后的弥生时代开始的。

織物か編物か?

紀元前3500年ころから始まる青銅器時代、世界各地の出土品に、経糸だけで織る、織物と編物の中間のような伸縮性を持つ、スプラング織(P.28)の布類があります。

Fabric or knit?
是织品还是编品?

In the Bronze Age, which began around 3500 BC, Sprang weaving (P.28) woven only with warps was found among excavated articles from various areas of the world. This fabric has elastic property between that of woven fabric or that of knitted item.

在大约公元前3500年开始的青铜器时代，从世界各地挖掘出仅由经线织成的Sprang(第28页)织。这种织品具有弹性，就像针织织物一样。

*経錦：中国で8世紀くらいまで織り続けられていた織り方。中国の戦国時代とは、紀元前400～200年くらい、日本では弥生時代の初めのころです。

*Vertical brocade : Weave that has been woven in China until about 8th century. China's Warring States Period was around 400 to 200 BC, when Japan was at the beginning of the Yayoi era.

*垂直锦缎：中国直至8世纪为止采用的编织方法。中国的战国时期大约在公元前400~200年，而日本则是在弥生时代初期。

織りと編み、重さ比べ / Weaving and knitting, weight comparison / 织与编的重量比较

糸から布を作りたい / Want to make cloth from yarn / 想从纱做成布

子どものころ、自分の手で布を作りたいと考え、母に相談しました。その答えは編み物の習得でした。それから60年以上編んでいます。

織物は2013年、『絹大好き 快適・健康・きれい』の出版後、取材した工房で習いました。

様々な絹糸でショールなどを織り、織物と同じ糸で、セーターなどを編みました。編み物用に多めに糸を買っても、足りないことがあります。

このページでは、同じ糸で同じサイズの織物と編み物を作り、使用量の違いを書きました。

When I was a child, I wanted to make cloth by myself with my own hands and consulted how to do to my mother.

Her answer was learning knitting. I have been knitting for more than 60 years since then.

I learnt about handwoven at the workshop where I interviewed in 2013 after the publication "Love silk Health: Comfortable, and Beautiful".

I wove a shawl with various silk yarns and knitted a sweater with the same yarn. I had an experience that yarn became short in case of knitting though I bought more volume of yarn for knitting than that for weaving. In this page, necessary yarn volume difference between woven fabric and knitted fabric was described, knitting and weaving of same size with the same thread and wrote the difference in usage.

童年时代，我咨询过母亲，因为我想用自己的双手做布。答案是针织类。从那以后长达60多年，我都用手工编织。

而织东西，2013 年则是在【挚爱丝绸 舒适•健康•美丽】出版之后，我在面試的作坊中学的。

我试着用各种各样的丝线织披肩，然后用同样的丝线编织毛衣。但即便是我囤积买了很多的丝线，还是感觉不够用。因此在这里，我用同样的丝线织成同样尺寸的织品以及编品，然后记录他们丝线用法差异。

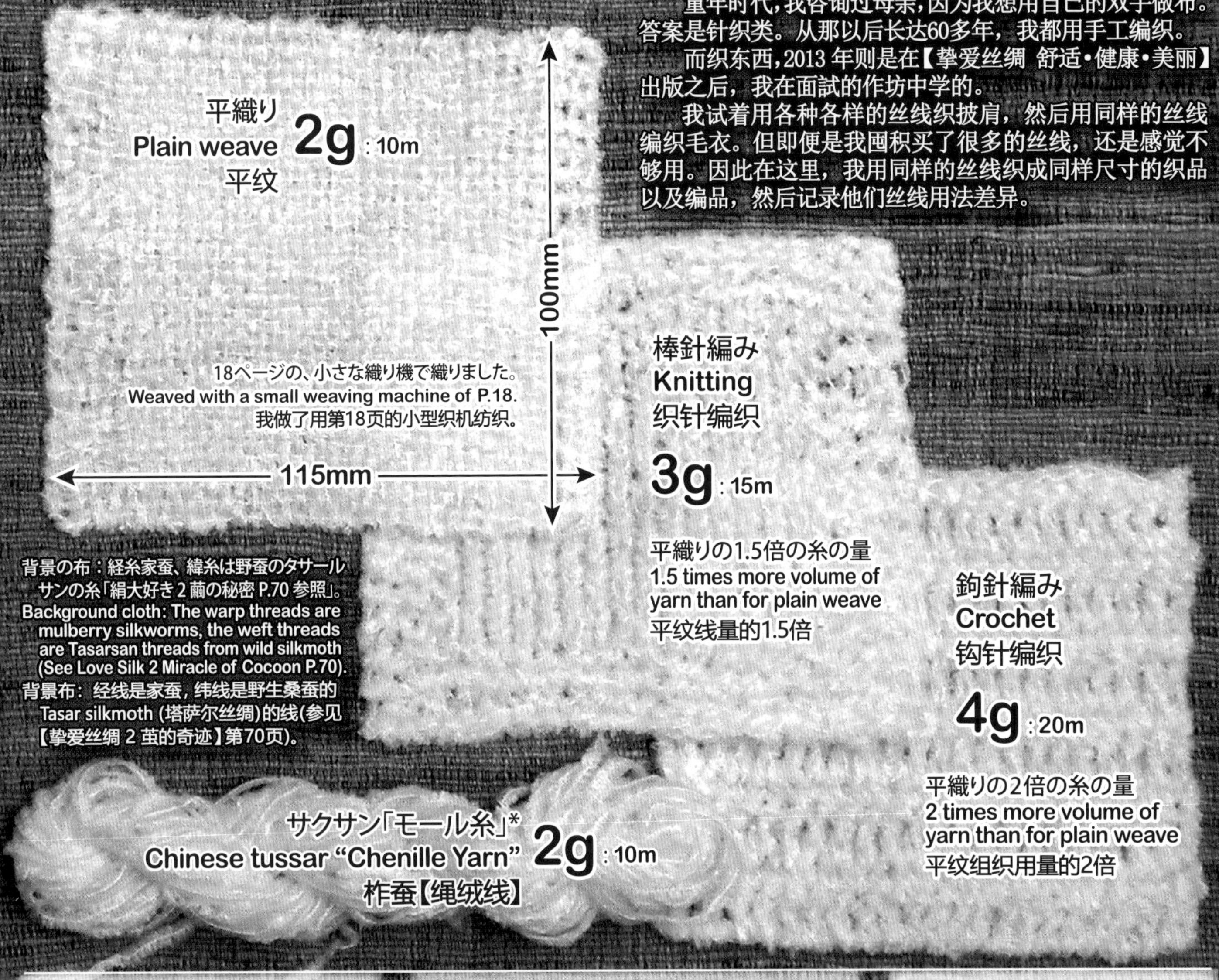

平織り
Plain weave **2g** : 10m
平纹

100mm

115mm

18ページの、小さな織り機で織りました。
Weaved with a small weaving machine of P.18.
我做了用第18页的小型织机纺织。

棒針編み
Knitting
织针编织

3g : 15m

平織りの1.5倍の糸の量
1.5 times more volume of yarn than for plain weave
平纹线量的1.5倍

鉤針編み
Crochet
钩针编织

4g : 20m

平織りの2倍の糸の量
2 times more volume of yarn than for plain weave
平纹组织用量的2倍

背景の布：経糸家蚕、緯糸は野蚕のタサールサンの糸「絹大好き2 繭の秘密 P.70 参照」。
Background cloth: The warp threads are mulberry silkworms, the weft threads are Tasarsan threads from wild silkmoth (See Love Silk 2 Miracle of Cocoon P.70).
背景布：经线是家蚕，纬线是野生桑蚕的Tasar silkmoth (塔萨尔丝绸) 的线(参见【挚爱丝绸 2 茧的奇迹】第70页)。

サクサン「モール糸」*
Chinese tussar "Chenille Yarn" **2g** : 10m
柞蚕【绳绒线】

*サクサン「モール糸」：野蚕のサクサン糸（絹大好き 2 繭の秘密 P.69 参照）を意匠撚糸機という特殊な撚糸機で撚ったファンシーヤーン（P.94）の一種。

*Chinese tussar "Chenille Yarn": A type of fancy yarn (P.94) which is made by wild silkmoth "Chinese oak silkmoth", twisted with a special Twisting machine called a Design twisting machine, (see "Love silk 2 Miracle of Cocoon" P.69).

*柞蚕「绳绒线」：由野生蚕吐丝制成的纱线——中国栎蚕蛾纱线(参见【挚爱丝绸2茧的秘密】第69页)——称为特殊的捻纱机，通过加捻的制成的一种纱线(参见94页)。

手織り体験 1. 腰織(こしばた) / Hand-weaving experience 1. Backstrap loom / 手工织体验 1. 腰织

指導：工藤 いづみ / Instructor: KUDO Izumi / 指导：工藤 Izumi

撮影場所：WILD SILK MUSEUM (P.117)、手織工房 SOX

Shooting location: WILD SILK MUSEUM (P.117), Hand-weaving studio SOX / 拍摄地点：野生丝绸博物馆 (第 117 页)，手织工房 SOX

「同じ幅で織るのって、緊張するわ!」
"Keeping the same width at weaving makes me nervous!"
【保持相同的宽度编织，使我感到紧张！】

*腰織に使う道具：工藤いづみ氏の「手織工房SOX」で、細い織物を織るために考案された「ベルト織用セット」を使用。
http://www.teorikoubousox.biz/

*Tools used for Backstrap loom: "Belt weaving set" designed to weave narrow fabrics at "Hand-weaving studio SOX" owned over by Ms. Kudo.
http://www.teorikoubousox.biz/index.html

*腰织使用的工具：在工藤女士的工作室【手织工房SOX】中，使用了专门织窄幅织品的【窄带织套件】。
http://www.teorikoubousox.biz/

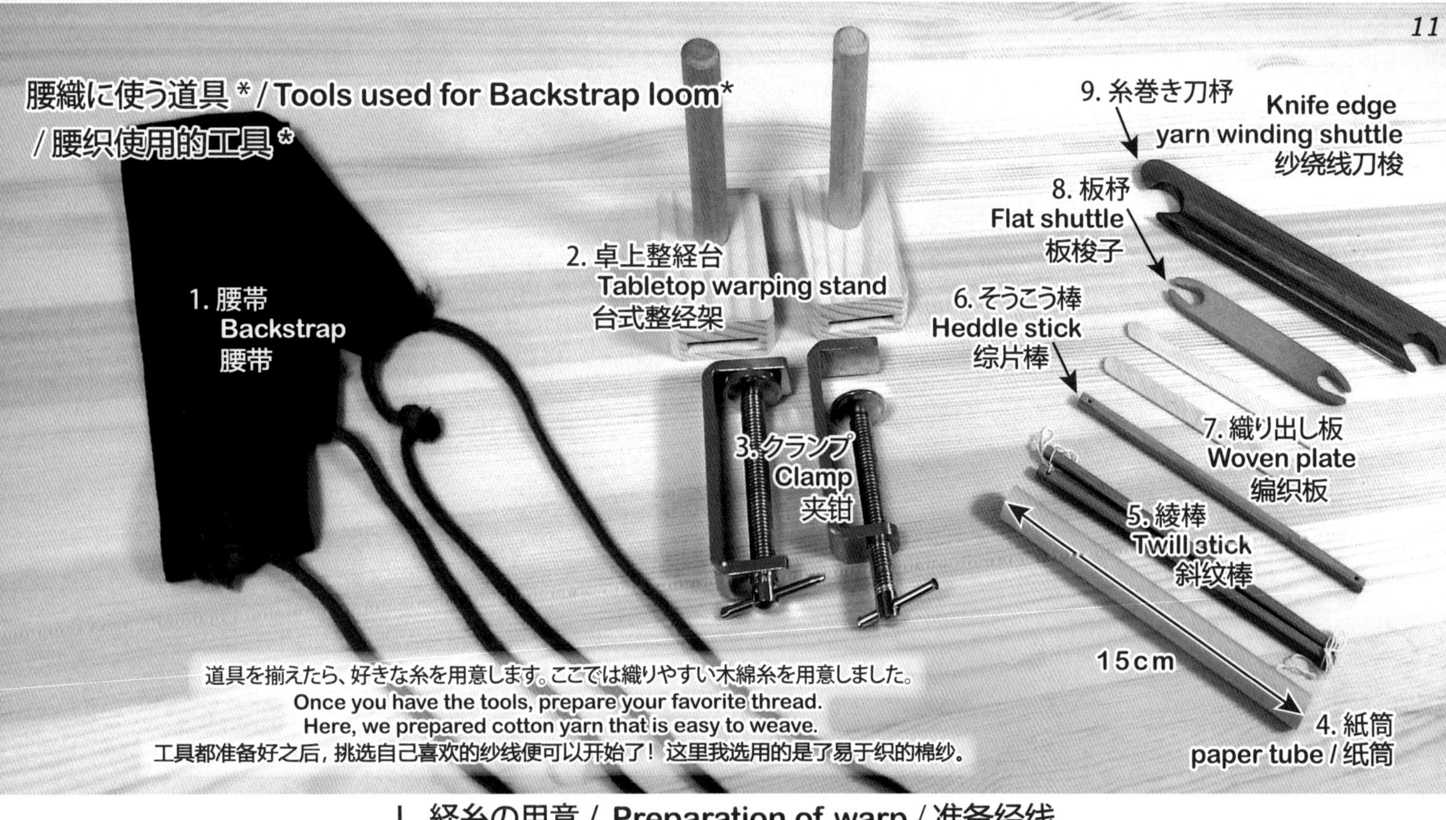

Ⅰ. 経糸の用意 / Preparation of warp / 准备经线

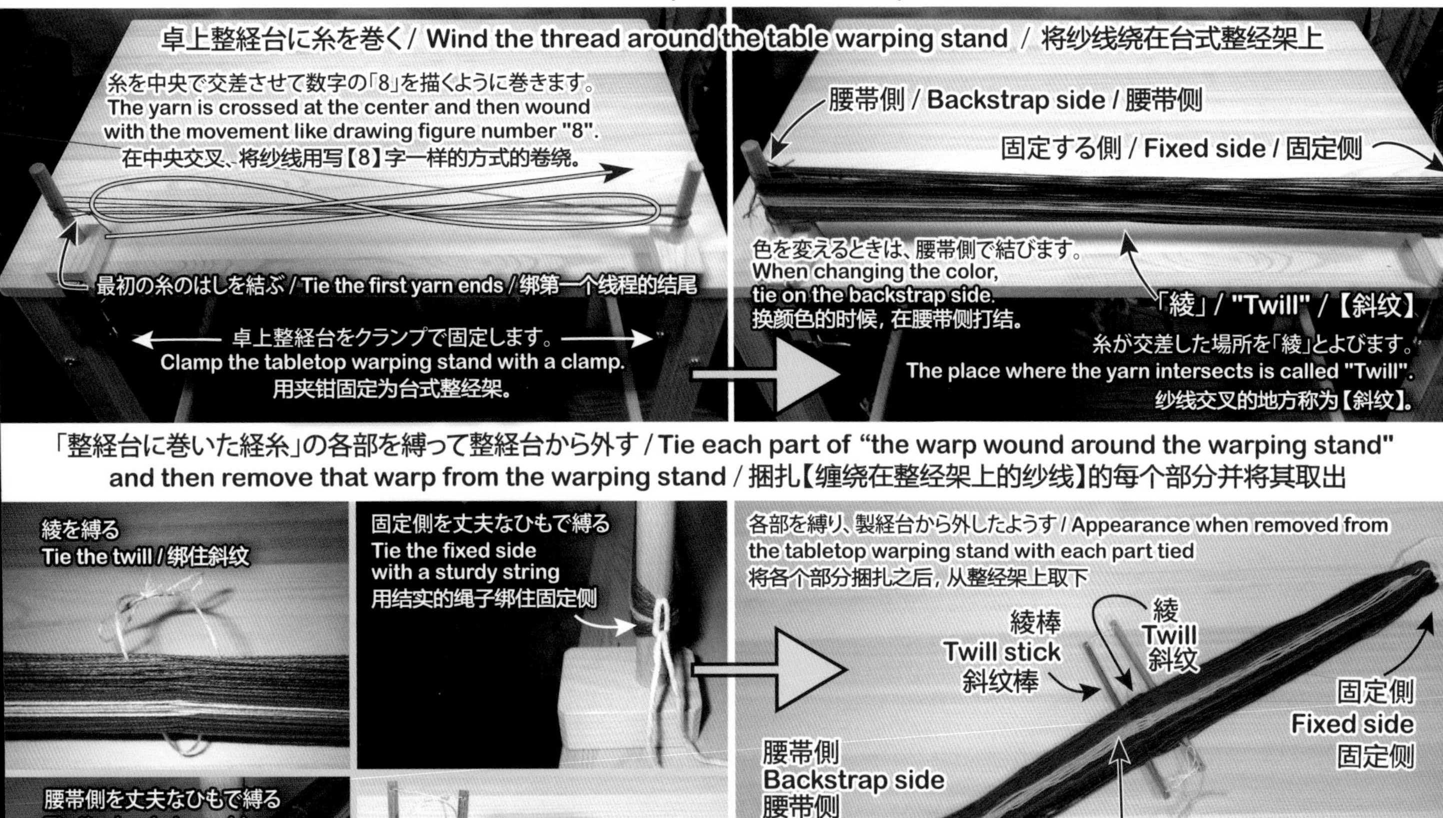

II. 経糸をテーブルと腰に固定する
Fix the warp to the table and the waist / 将经线固定到桌子以及自己的腰部

固定側のひもをクランプでテーブルに固定します。
Fix the string of the fixed side to the table with a clamp.
将固定侧的绳子用夹钳固定到桌子上。

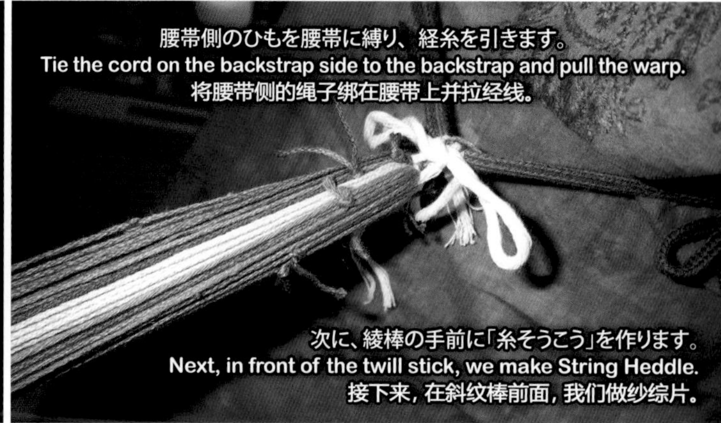

腰帯側のひもを腰帯に縛り、経糸を引きます。
Tie the cord on the backstrap side to the backstrap and pull the warp.
将腰带侧的绳子绑在腰带上并拉经线。

次に、綾棒の手前に「糸そうこう」を作ります。
Next, in front of the twill stick, we make String Heddle.
接下来，在斜纹棒前面，我们做纱综片。

III. 糸そうこうの作り方 / How to make a String Heddle / 纱综片的制作方法

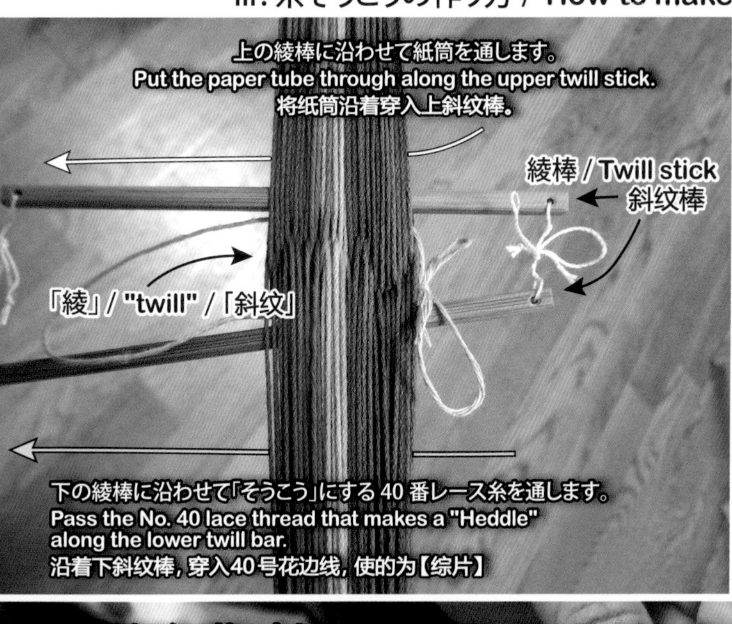

上の綾棒に沿わせて紙筒を通します。
Put the paper tube through along the upper twill stick.
将纸筒沿着穿入上斜纹棒。

下の綾棒に沿わせて「そうこう」にする 40 番レース糸を通します。
Pass the No. 40 lace thread that makes a "Heddle" along the lower twill bar.
沿着下斜纹棒，穿入40号花边线，使的为【综片】

横から見たところ / Seen from the side / 从侧面看

そうこう棒に 40 番レース糸で、すべての下糸を左端から順番に拾います。
On top of the Heddle stick, pick up all the lower threads with the No. 40 lace thread in order from the left end.
在综片棒的顶部，从左端开始按40号花边线挑起所有底线。

下側の糸の拾い方を上から見る / See how to pick up the lower thread from above / 如何拾取从上方查看的底线程

下側の糸の拾い方
How to pick up lower thread / 如何捡线起底线

そうこう棒の両端を結ぶ / Tie both ends of heddle stick
将综片棒的两端打结

下側の糸
Lower thread
底线

糸そうこうの完成
Completion of string heddle / 纱综片完成

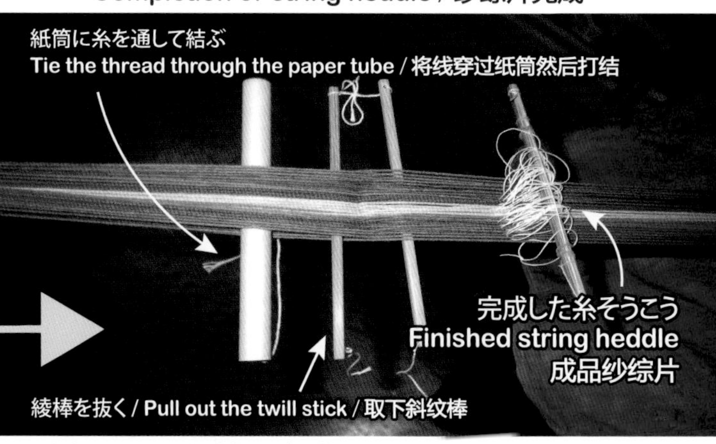

Ⅳ.「織り出し板」を経糸のベルト側に通して腰機の完成!

Pass the "woven plate" to the Backstrap side of the warp to complete the Backstrap loom !
将【编织板】穿过经线的腰带侧，一台腰织机就完成了！

経糸を腰で強く引くことが重要
It's important to pull the warp strongly with waist
重要腰部发力将经线拉紧很

織り出し板
Woven plate
编织板

織り始める前に、緯糸を糸巻き刀杼に巻きます。
Before weaving, wind weft around knife edge yarn winding shuttle.
开始织之前，将纬线缠绕到纱绕线刀梭上。

Ⅴ. いよいよ織り始め / Finally weaving start / 可以开始织制

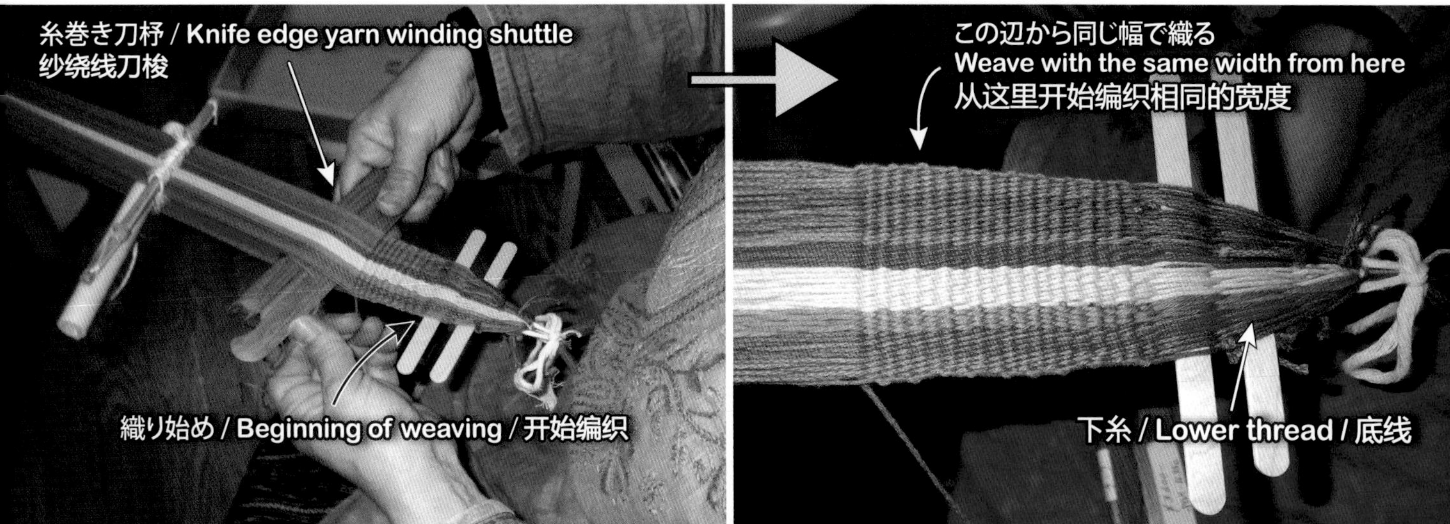

VI. 織の途中で固定部から外す
Unfix from the fixed part during weaving / 可在织制中途，从固定部分拆下

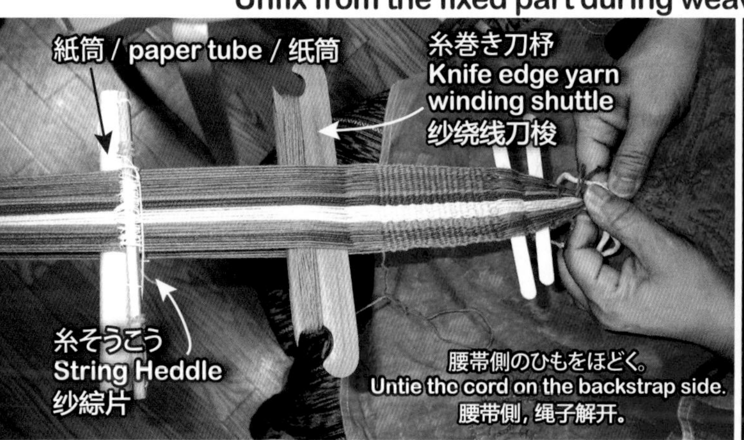

取り外し可能な腰機は、どこでも織れる
With removable Backstrap loom, weaving is possible at any location / 可以携带的腰织机，在哪里都可以织制

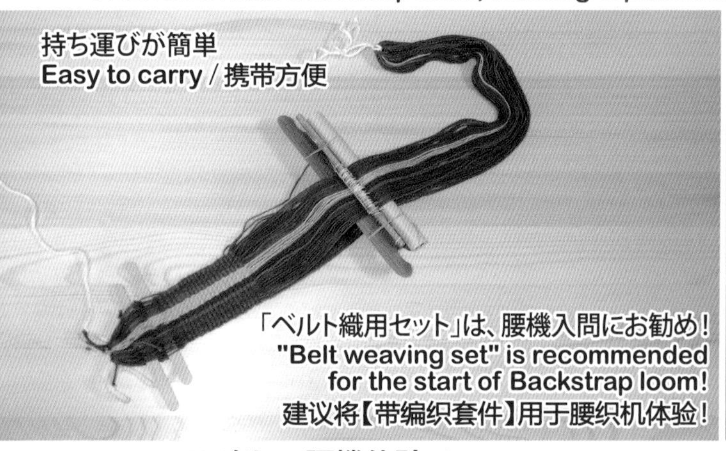

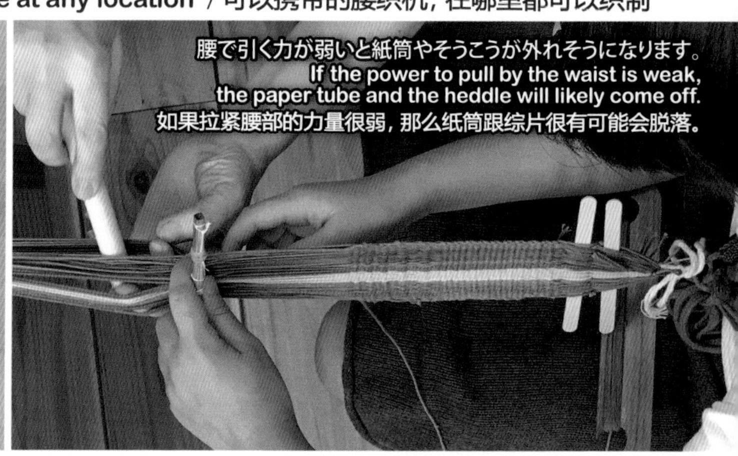

VII. 楽しい腰機体験 / Happy experience of Backstrap loom / 有趣的腰织机体验

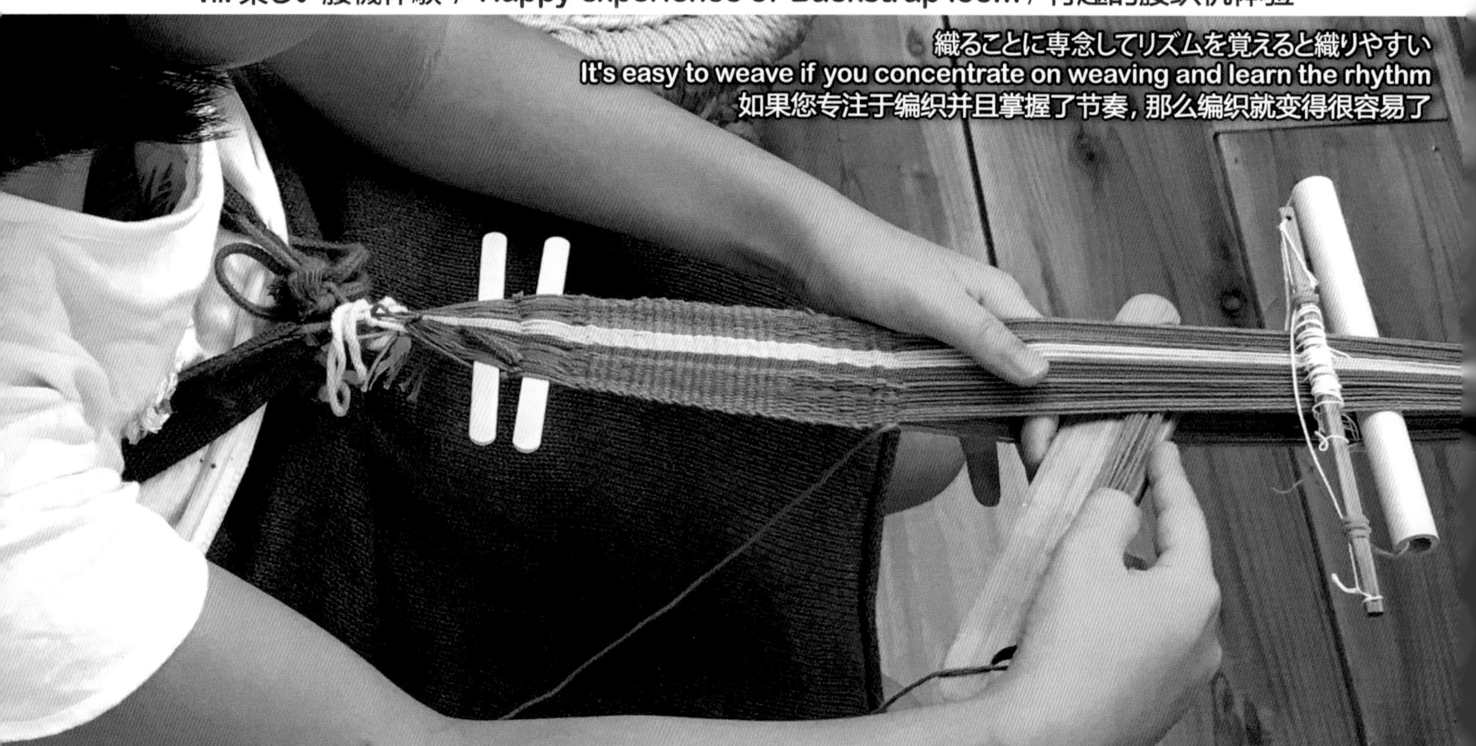

手織り体験 2. 腰織の理論で細いリボンを織る
Hand-weaving experience 2. Weave a narrow ribbon applying Back strap loom theory
手工织体验 2. 用腰织机的原理，编织一条窄幅缎带

身近な材料を使って織機を作る / Making a loom with everyday materials / 使用熟悉的材料作成织机

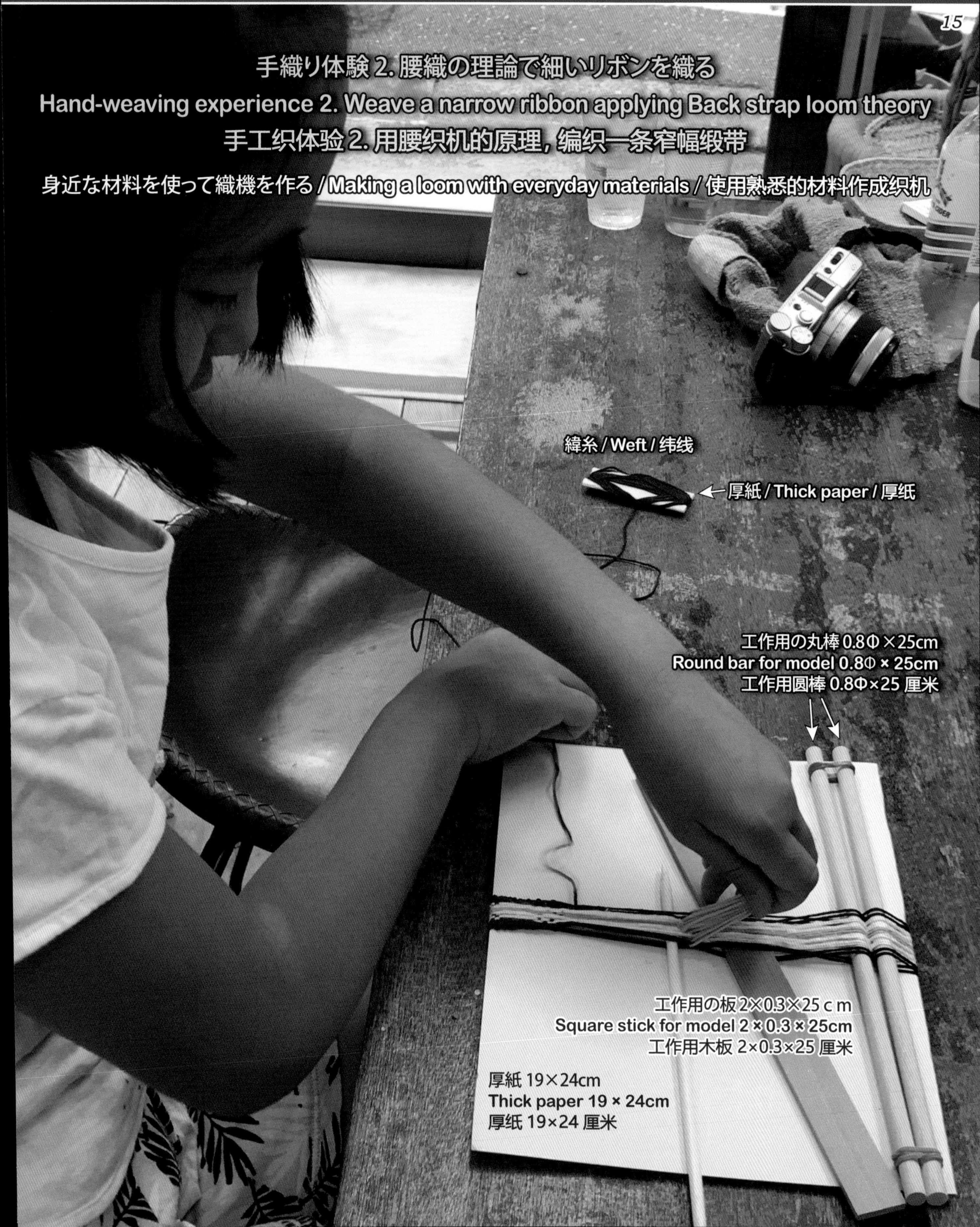

緯糸 / Weft / 纬线

厚紙 / Thick paper / 厚纸

工作用の丸棒 0.8Φ×25cm
Round bar for model 0.8Φ × 25cm
工作用圆棒 0.8Φ×25 厘米

工作用の板 2×0.3×25 c m
Square stick for model 2 × 0.3 × 25cm
工作用木板 2×0.3×25 厘米

厚紙 19×24cm
Thick paper 19 × 24cm
厚纸 19×24 厘米

I. 厚紙に経糸(絹)を巻く / Wrap warp (silk) on thick paper / 将经线(丝绸)滚缠绕在厚纸上

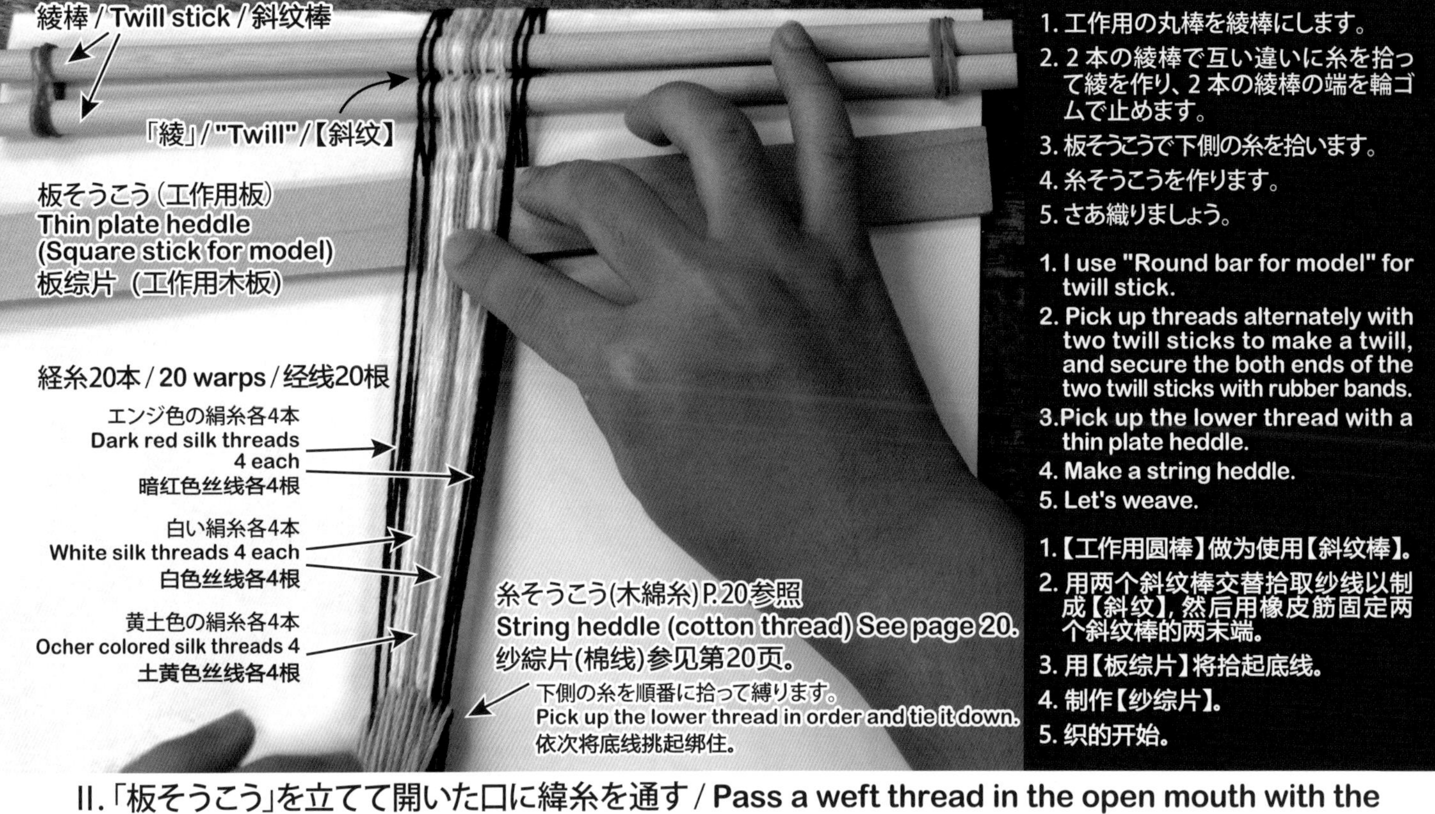

1. 工作用の丸棒を綾棒にします。
2. 2本の綾棒で互い違いに糸を拾って綾を作り、2本の綾棒の端を輪ゴムで止めます。
3. 板そうこうで下側の糸を拾います。
4. 糸そうこうを作ります。
5. さあ織りましょう。

1. I use "Round bar for model" for twill stick.
2. Pick up threads alternately with two twill sticks to make a twill, and secure the both ends of the two twill sticks with rubber bands.
3. Pick up the lower thread with a thin plate heddle.
4. Make a string heddle.
5. Let's weave.

1. 【工作用圆棒】做为使用【斜纹棒】。
2. 用两个斜纹棒交替拾取纱线以制成【斜纹】, 然后用橡皮筋固定两个斜纹棒的两末端。
3. 用【板综片】将拾起底线。
4. 制作【纱综片】。
5. 织的开始。

II. 「板そうこう」を立てて開いた口に緯糸を通す / Pass a weft thread in the open mouth with the "thin plate heddle" upright / 使【板综片】直立, 将纬纱穿过开口处

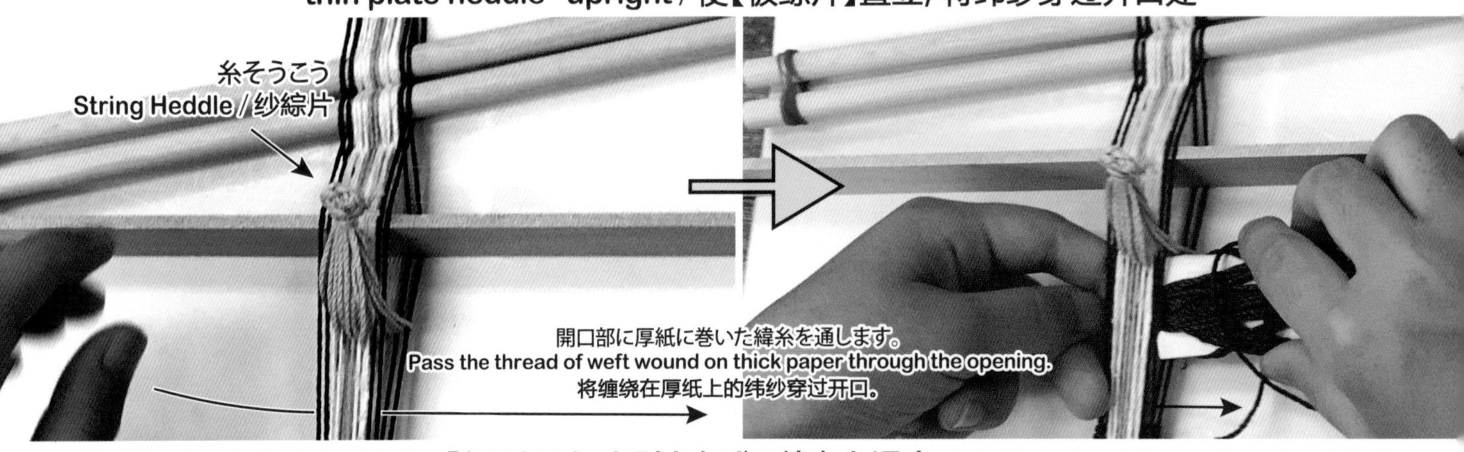

III. 「板そうこう」を平らにし、「糸そうこう」を引き上げて緯糸を通す / Flatten the "thin plate heddle", pull up the "string heddle" and pass the weft / 将【板综片】放平, 拉起【纱综片】让纬线穿过

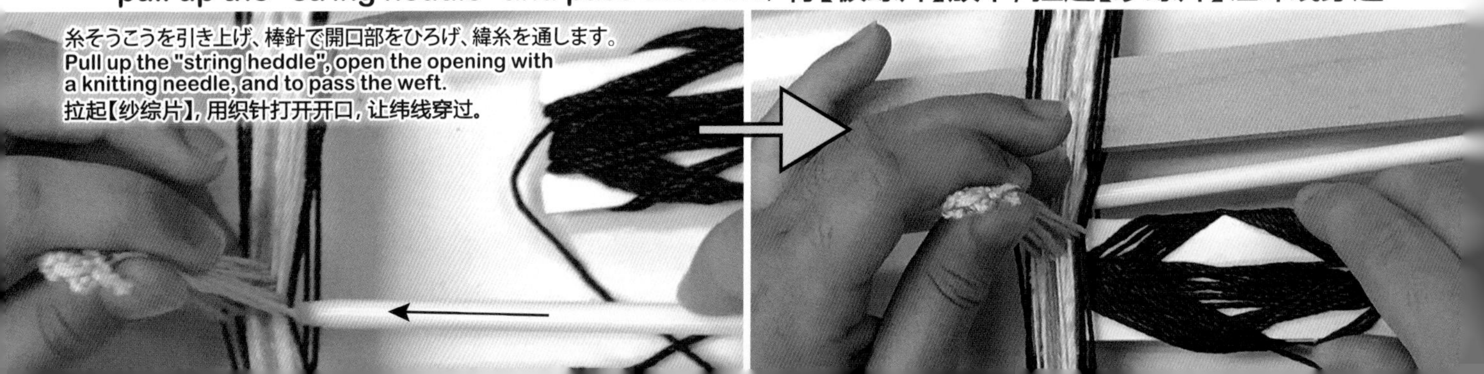

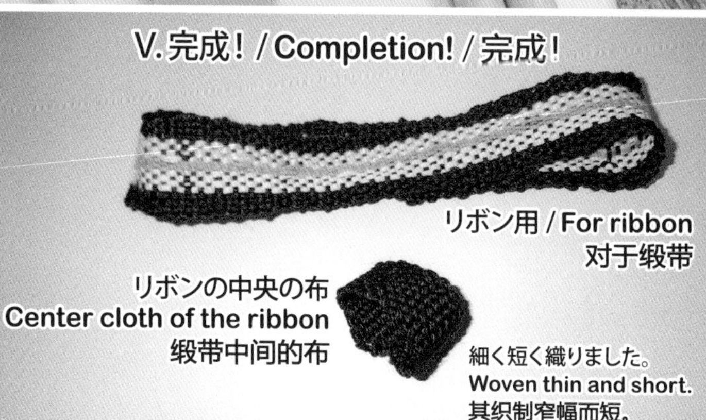

VI. 世界にただ一つの手織りシルクリボン
Only one hand-woven silk ribbon in the world / 世界上唯一手工编织的丝绸缎带

手織り体験3. 腰機で小さな四角い布を織る
Hand-weaving experience 3. Weave a small square cloth with a Backstrap loom
手工织体验 3. 用腰织机织制一块小方布

アートヤーン*とファンシーヤーンのミニ織（実際の大きさ）
Mini weaving of art yarn* and fancy yarn (actual size) / 艺术纱*与设计加捻纱的简易迷你织制（实际尺寸）

緯糸：インド製アートヤーン(絹)
Weft : Indian made art yarn (silk)
纬线：印度产的艺术纱(丝绸)

サクサン「モール糸」
Chinese tussar "Chenille yarn"
柞蚕「绳绒线」

*アートヤーン（絹手紡ぎ）：上段中央の経糸と緯糸、下段左右の緯糸。工房風花（絹遊塾）P.117。

*Art yarn (silk hand-spun) : Warp and weft of upper center, and weft of lower left and right. Atelier Kazahana (silk school) P.117.

*艺术纱(真丝手纺)：上段中间的布料为经纱与纬纱。下段左右布料为纬线。工房Kazahana (絹遊塾) 第117頁。

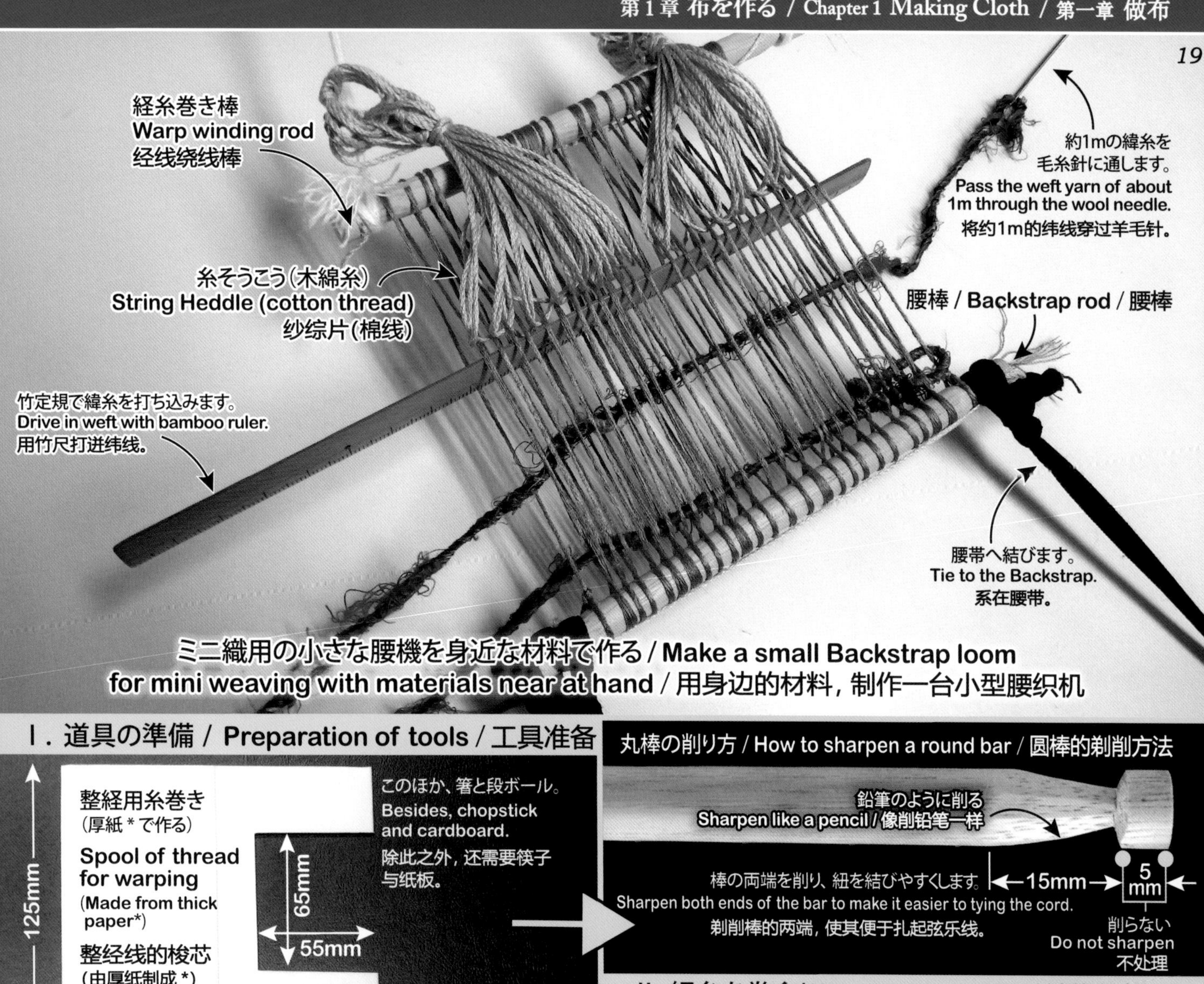

ミニ織用の小さな腰機を身近な材料で作る / Make a small Backstrap loom for mini weaving with materials near at hand / 用身边的材料，制作一台小型腰织机

Ⅰ. 道具の準備 / Preparation of tools / 工具准备

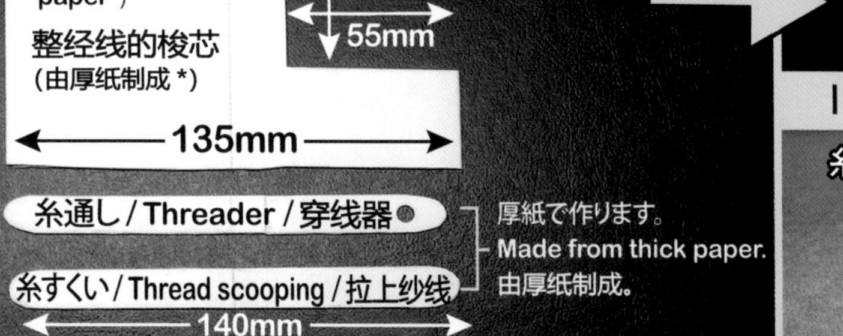

このほか、箸と段ボール。
Besides, chopstick and cardboard.
除此之外，还需要筷子与纸板。

糸通し / Threader / 穿线器
糸すくい / Thread scooping / 拉上纱线
140mm
厚紙で作ります。
Made from thick paper.
由厚纸制成。

経糸巻き棒 / Warp winding rod / 经线绕线棒
腰棒 / Backstrap rod / 腰棒
丸棒（直径8mm）
Round bar (diameter 8 mm)
圆棒(直径8毫米)

竹定規 20cm / Bamboo ruler 20cm / 竹尺20厘米

棒針（糸すくい）/ Knitting needle (Thread scooping) / 织针(拉上纱线)

カギ針（糸すくい）/ Crochet needle (Thread scooping) / 钩针(拉上纱线)

毛糸針 —— Wool needle / 羊毛针

丸棒の削り方 / How to sharpen a round bar / 圆棒的剃削方法

鉛筆のように削る
Sharpen like a pencil / 像削铅笔一样

棒の両端を削り、紐を結びやすくします。
Sharpen both ends of the bar to make it easier to tying the cord.
剃削棒的两端，使其便于扎起弦乐线。

15mm
5 mm
削らない
Do not sharpen
不处理

Ⅱ. 経糸を巻く / Winding the warp / 缠绕经线

糸の巻き方 / How to wind the thread / 经纱的缠绕方法

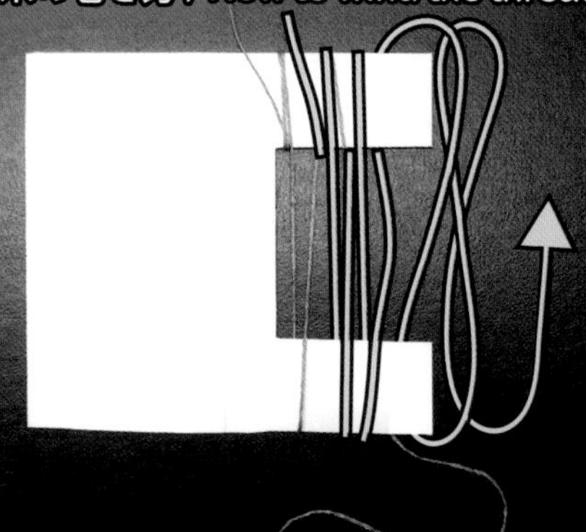

糸を前から後ろ、前から後ろへ掛けて、糸を中央で交差させて数字の「8」を描くように巻きます。

Hang the thread repeatedly from front to back and after crossing it at the center, wind it with the movement like drawing figure number "8".

将线从前到后悬挂，从前到后悬挂，使纱线在中心交叉并卷绕以绘制【8】数字。

経糸：絹紡糸 (P.49) 6m
Warp : silk spun yarn (P.49) 6 m / 经纱：绢丝(第49页)6m

＊厚紙：牛乳パックを二つ折りにして、糊で貼り付けて使用しています。約10mm幅の糸通しや、糸すくいも、ていねいに切り抜きましょう。

＊Thick paper: Milk carton is half folded and pasted with glue. Let us carefully cut out "threader" of about 10 mm width and "thread scooping".

＊厚纸：这里我们将牛奶盒两折叠起来，然后用胶水粘住使用。然后小心地划开大约10毫米宽的【穿线器】和【拉上纱线】。

III. 綾を固定する / Fix the twill / 固定斜纹

IV. 織り機を組み立てる / Assembling the loom / 织机的组装

V.「糸そうこう」を作る Making "String Heddle" / 制作【纱综片】

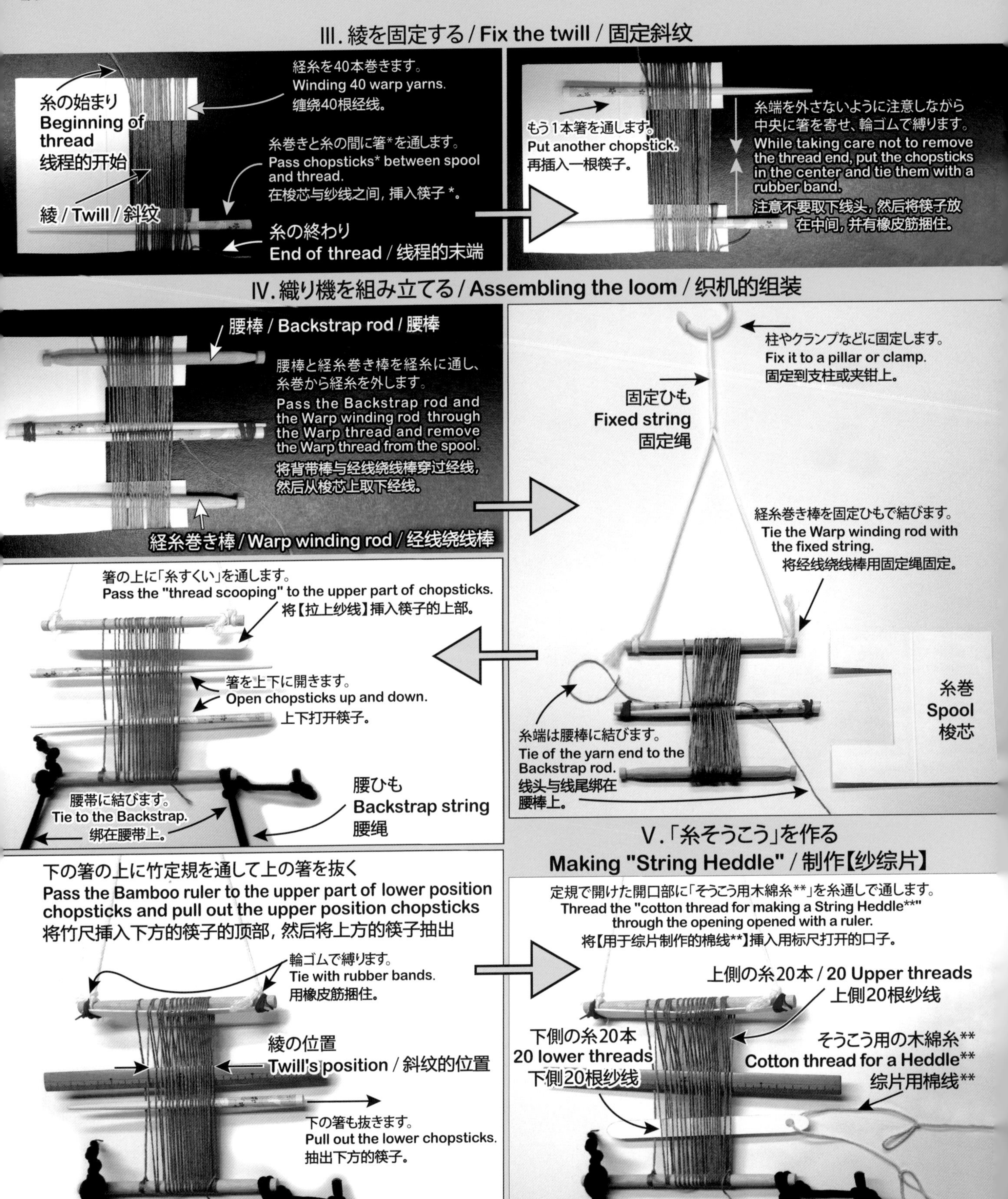

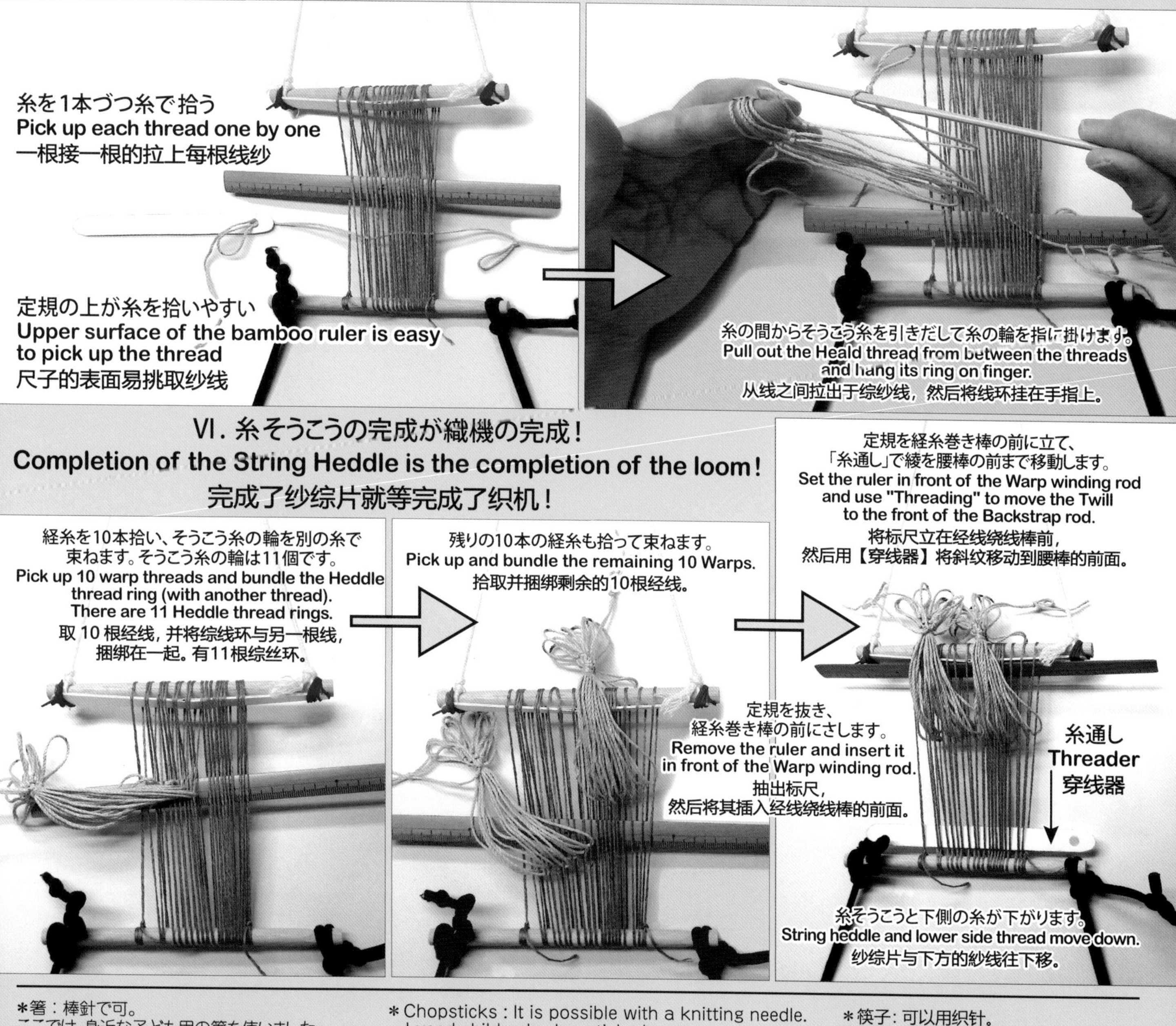

＊箸：棒針で可。
ここでは、身近な子ども用の箸を使いました。

＊Chopsticks : It is possible with a knitting needle.
I used children's chopsticks here.

＊筷子：可以用织针。
这里我使用了的儿童筷子。

**糸そうこう用木綿糸の作り方。/ How to make cotton thread for a String Heddle. / 纱综片用棉线的制作方法。

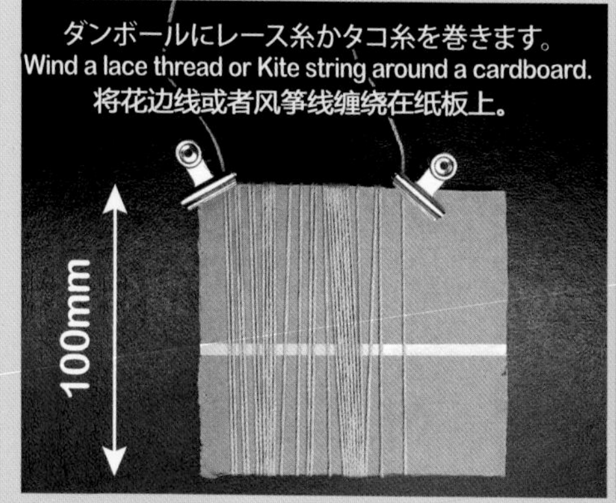

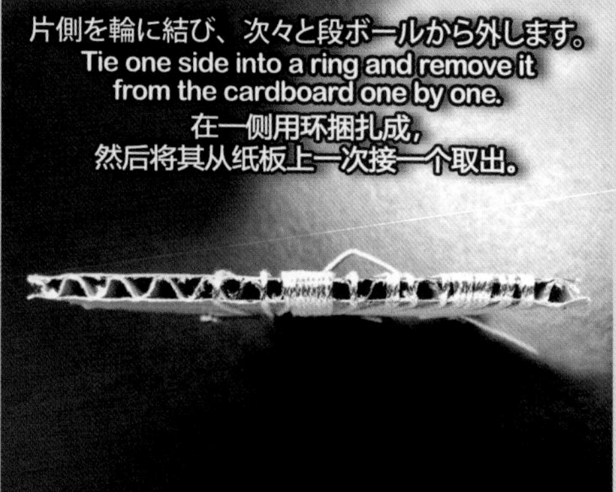

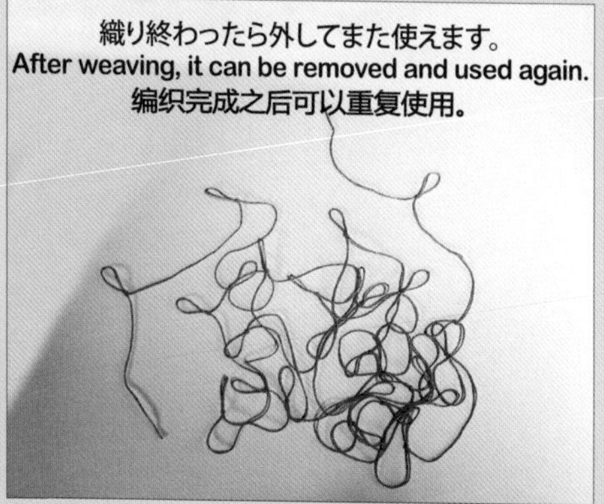

Ⅶ. 織り始め / Start of weaving / 开始编织

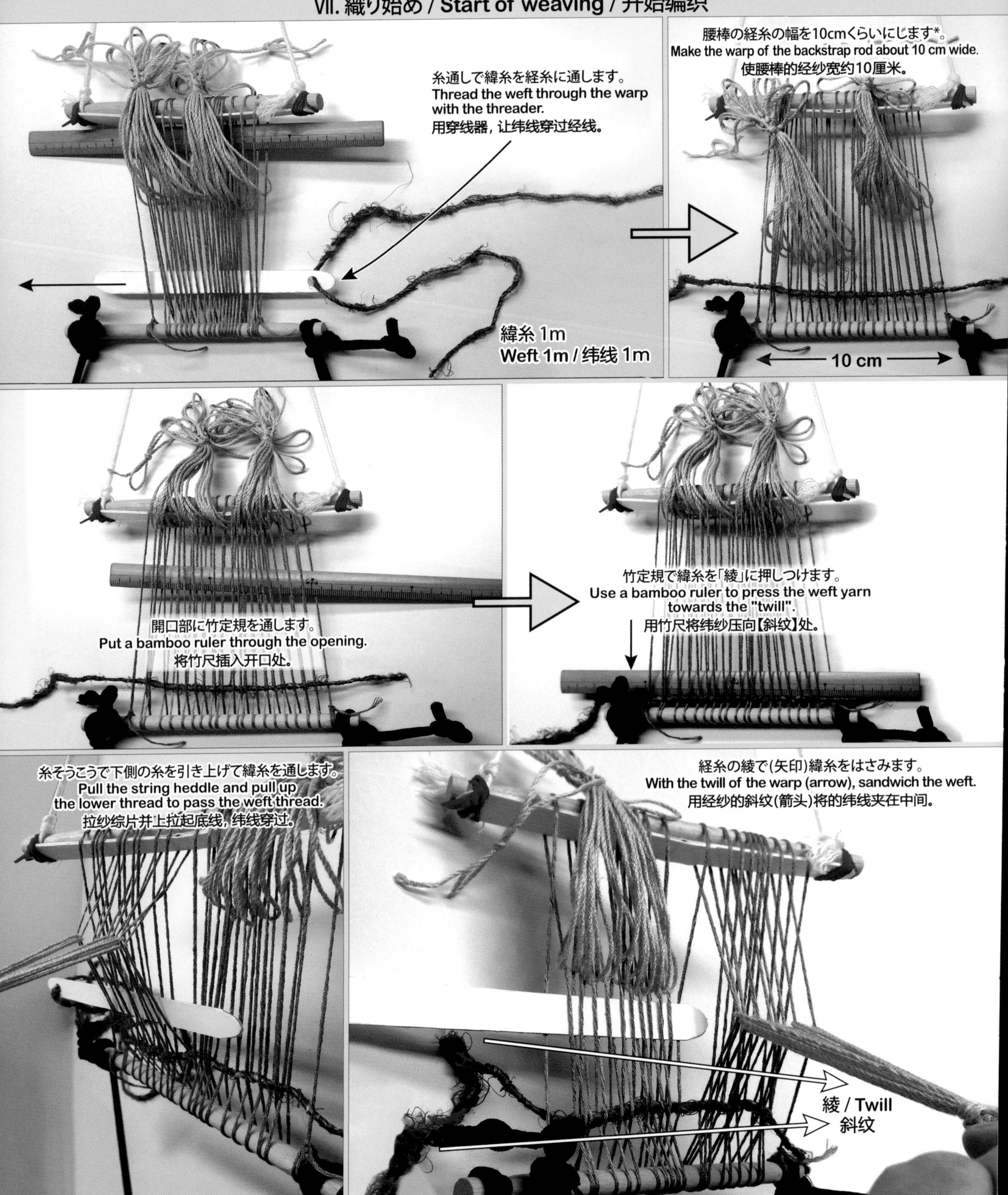

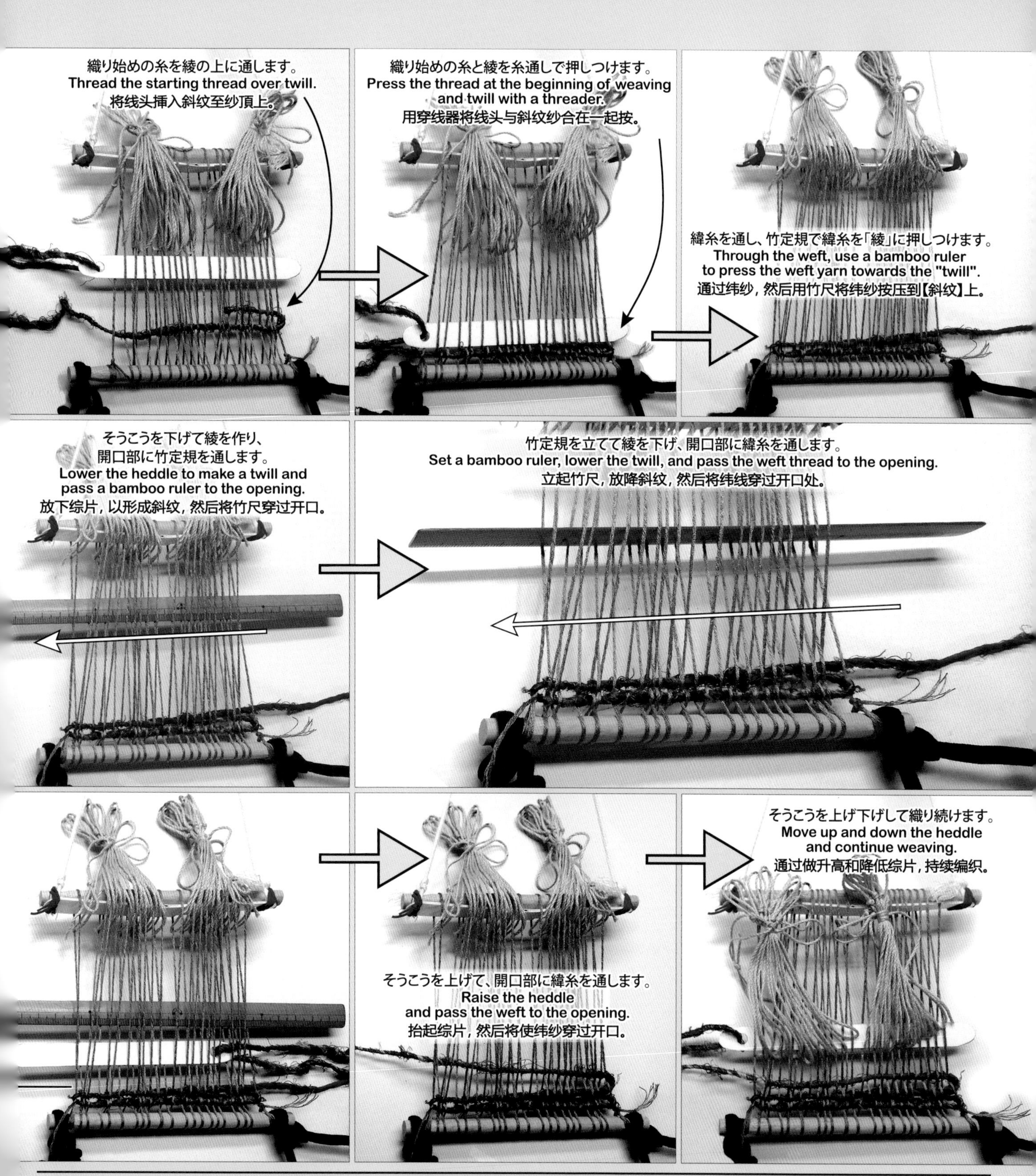

＊10cmくらい：織りたい幅に合わせ、経糸巻き棒と腰棒の長さを選択します。

* Approximately 10 cm: Select the length of the warp winding rod and waist rod according to the width you want to weave.

＊约10厘米：根据要编织的宽度，选择经线绕线棒与腰棒的长度。

Ⅷ. 織り終わり
End of weaving / 织造结束

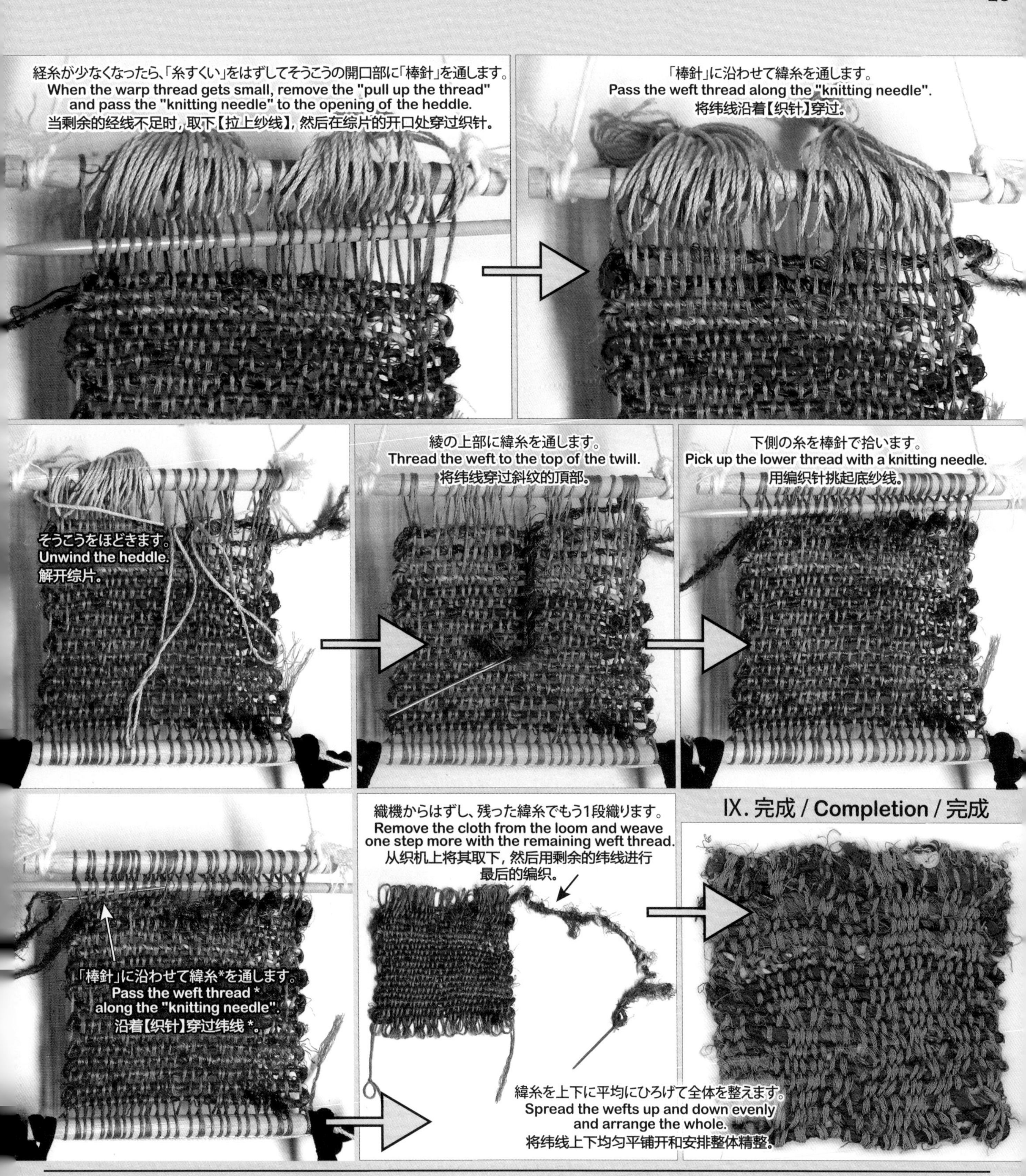

Ⅸ. 完成 / Completion / 完成

*緯糸：ここでは緯糸を毛糸針で通しましたが、左ページのように畳針や長いゴム通しなど、通しやすい道具を使ってください。

*Wefts: Wefts are threaded here with a wool needle, but use a tool that is easy tothread, such as a tatami needle or a long Rubber threader, as shown on the left page.

*纬线：这里纬线是用羊毛针的，根据自己情况，可以用其他工具，比如榻榻米针或者长橡胶穿线器等，参見如左页所。

X. 織りはしの仕上げ方 / How to finish the edge of weaving / 编织的边缘收尾的方法

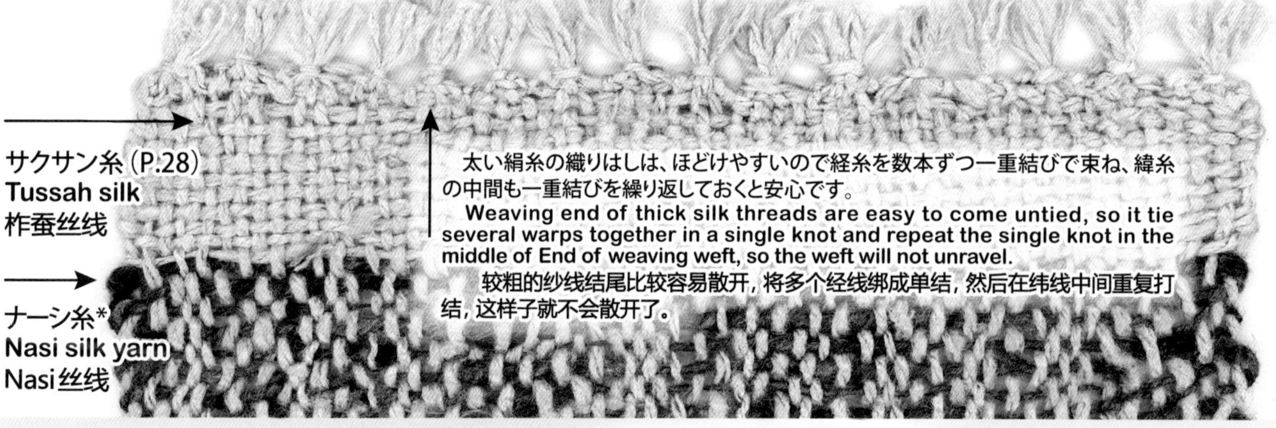

サクサン糸 (P.28)
Tussah silk
柞蚕丝线

ナーシ糸*
Nasi silk yarn
Nasi丝线

太い絹糸の織りはしは、ほどけやすいので経糸を数本ずつ一重結びで束ね、緯糸の中間も一重結びを繰り返しておくと安心です。
Weaving end of thick silk threads are easy to come untied, so it tie several warps together in a single knot and repeat the single knot in the middle of End of weaving weft, so the weft will not unravel.
较粗的纱线结尾比较容易散开，将多个经线绑成单结，然后在纬线中间重复打结，这样子就不会散开了。

サイドテーブルの脚で経糸を巻き、経糸巻き棒と腰棒を 25cm にした腰機で織った布
Woven cloth by winding warp threads on the legs of the side table and changing the warp winding rod and waist rod at a length of 25 cm
在侧桌腿上缠绕经线，然后将经线绕线棒与腰棒长的长度调改为25厘米的织布

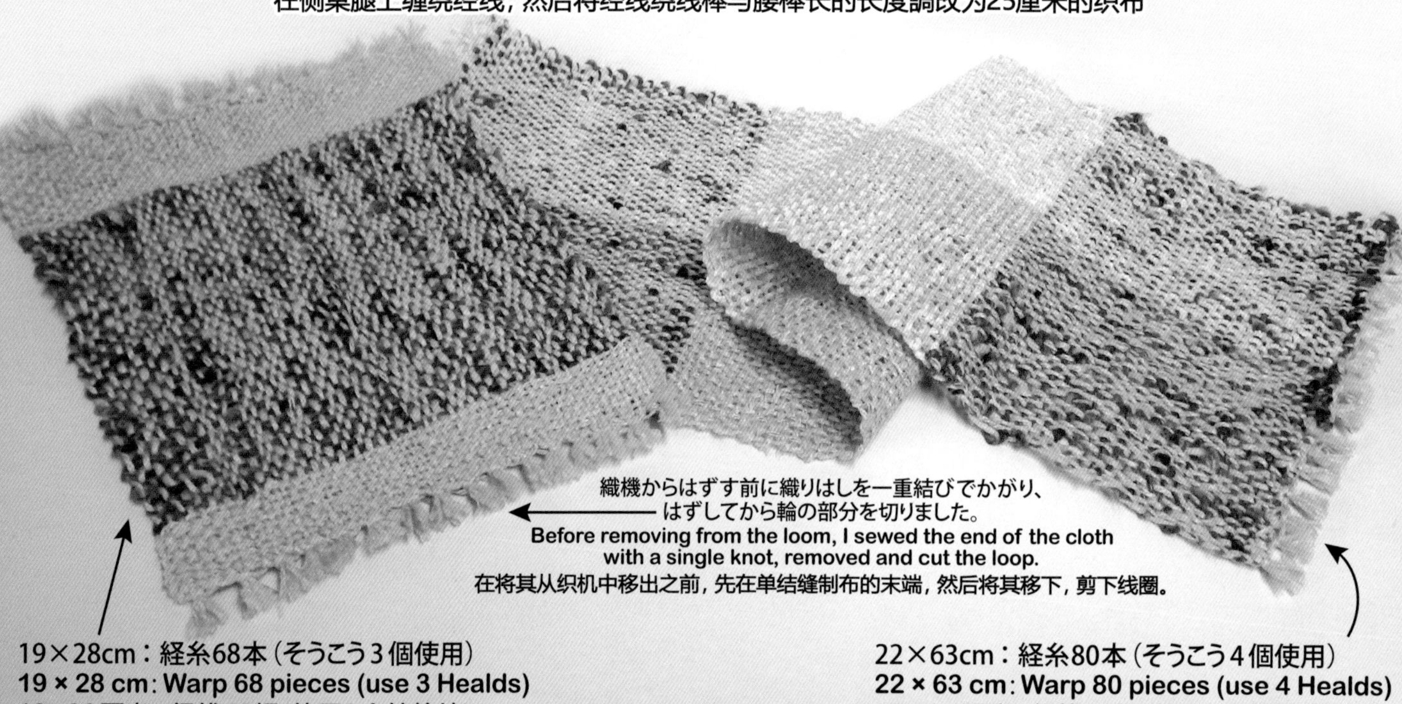

織機からはずす前に織りはしを一重結びでかがり、はずしてから輪の部分を切りました。
Before removing from the loom, I sewed the end of the cloth with a single knot, removed and cut the loop.
在将其从织机中移出之前，先在单结缝制布的末端，然后将其移下，剪下线圈。

19×28cm：経糸68本（そうこう3個使用）
19 × 28 cm: Warp 68 pieces (use 3 Healds)
19×28厘米：经线68根(使用3个纱综片)

22×63cm：経糸80本（そうこう4個使用）
22 × 63 cm: Warp 80 pieces (use 4 Healds)
22×63 厘米：经线80根(使用4个纱综片)

裏側から見た織りはしを一重結びで仕上げた様子。
Backside view of the fabric end being finished with a single knot.
从背面查看布料的末端如何打单结。

---

*ナーシ糸：アトリエ トレビ（P.134）
** 作品、*** 精練したキビソ糸を使用、糸：工房 風花（絹遊塾）（P.117）
**** ショール：八丁ヤーン㈱（P.116）

*Nasi thread: Atelier Trevi (P.134)
** Works, *** Using kibiso degummed silk yarn: Atelier Kazahana (silk school) (P.117)
**** Shawl: Hatcho Yarn Co., Ltd (P.116)

*Nasi线程：Atelier 特雷维 第134页
作品、* 使用精练的kibiso真丝纱线：工房 Kazahana (丝绸学校) 第117 页
**** 披肩：八丁纱线株式会社 第116 页

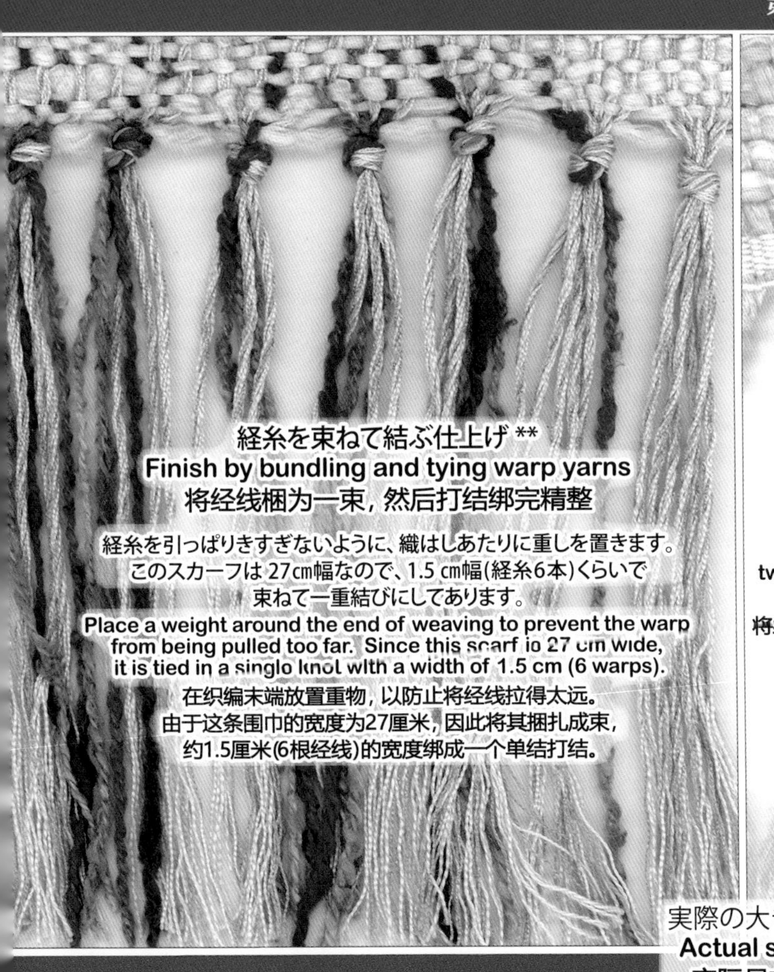

経糸を束ねて結ぶ仕上げ **
Finish by bundling and tying warp yarns
将经线捆为一束，然后打结绑完精整

経糸を引っぱりきすぎないように、織はしあたりに重しを置きます。
このスカーフは 27㎝幅なので、1.5 ㎝幅(経糸6本)くらいで
束ねて一重結びにしてあります。

Place a weight around the end of weaving to prevent the warp from being pulled too far. Since this scarf is 27 cm wide, it is tied in a single knot with a width of 1.5 cm (6 warps).

在织编末端放置重物，以防止将经线拉得太远。
由于这条围巾的宽度为27厘米，因此将其捆扎成束，
约1.5厘米(6根经线)的宽度绑成一个单结打结。

経糸をより合わせて房にする ***
Twist the Warp into a a tassel
将经线扭成流苏

経糸6本、約1.5㎝幅の糸を2つに分けて、それぞれを同じ方向に20回
よります。よった2本の糸を1本に合わせて反対方向に
10回くらいより、糸のはしを一重結びにします。

Divide the yarn of about 1.5 cm width (about 6 warps) into two and twist each in the same direction 20 times. Twist the two bundle of threads in the opposite direction about 10 times, twist them together and tie the ends of the threads into a single knot.

将约1.5厘米宽（约6根经纱）的纱线分成两部分，每根在相同方向上加捻20次；
然后将两根加捻线合二为一，沿相反方向再加捻10次，
然后在纱线的加捻末绑成一个单结。

実際の大きさ
Actual size
实际尺寸

美しい日本の着物の生地の布端を房にしてショールにする
Make a shawl with the end of a beautiful Japanese kimono fabric as a tuft
用美丽的日本和服面料制成披肩，使布的边缘簇绒。

この生地は洋装用の桐生お召しです。経糸を6㎜幅くらいとって
（約 80 本）二つに分け、同じ方向に 40 回ぐらいよってから
2 本を合わせて反対方向によって最後に結びます。

This fabric is Kiryu silk crepe weave for Western clothing. Take a warp about 6 mm wide(about 80 pieces), divide it into two parts, and twist it about 40 times in the same direction, then merge the two, twist it in the opposite direction, and then tie at the end.

这种面料于西服的Kiryu熟丝绉绸。取约6毫米宽(约80根)
的经线，将其分成两部分 并在相同方向上加捻40次，
将两者合二为一，并在相反方向上加捻，然后单结。

インドのサリーの布（絹）でショールを作る
Making shawls with Indian saree cloth (silk)
用印度纱丽布面(丝绸)和制作披肩

布端にロックミシンをかけます。織糸を切らないように注意しながら
5mmおきに目打ちで穴をあけてかぎ針で縁編をします。

Sew the cloth edge with a lock sewing machine. While taking care not to cut the weaving thread, perforate holes every 5 mm with a eyeleteer and knit the rim with a crochet.

用锁式缝纫机缝制布料边缘。注意不要割断织线，
用锥梃每 5 毫米打一个孔，并用钩针编织在边缘。

絹紡糸 / Spun silk yarn / 绢丝

手織り体験 4. 古代の技法「スプラング織」を楽しむ
Hand-weaving experience 4. Enjoy ancient techniques, "Sprang weave"
手工编织体验 4. 享受古老技巧【Sprang 编织】

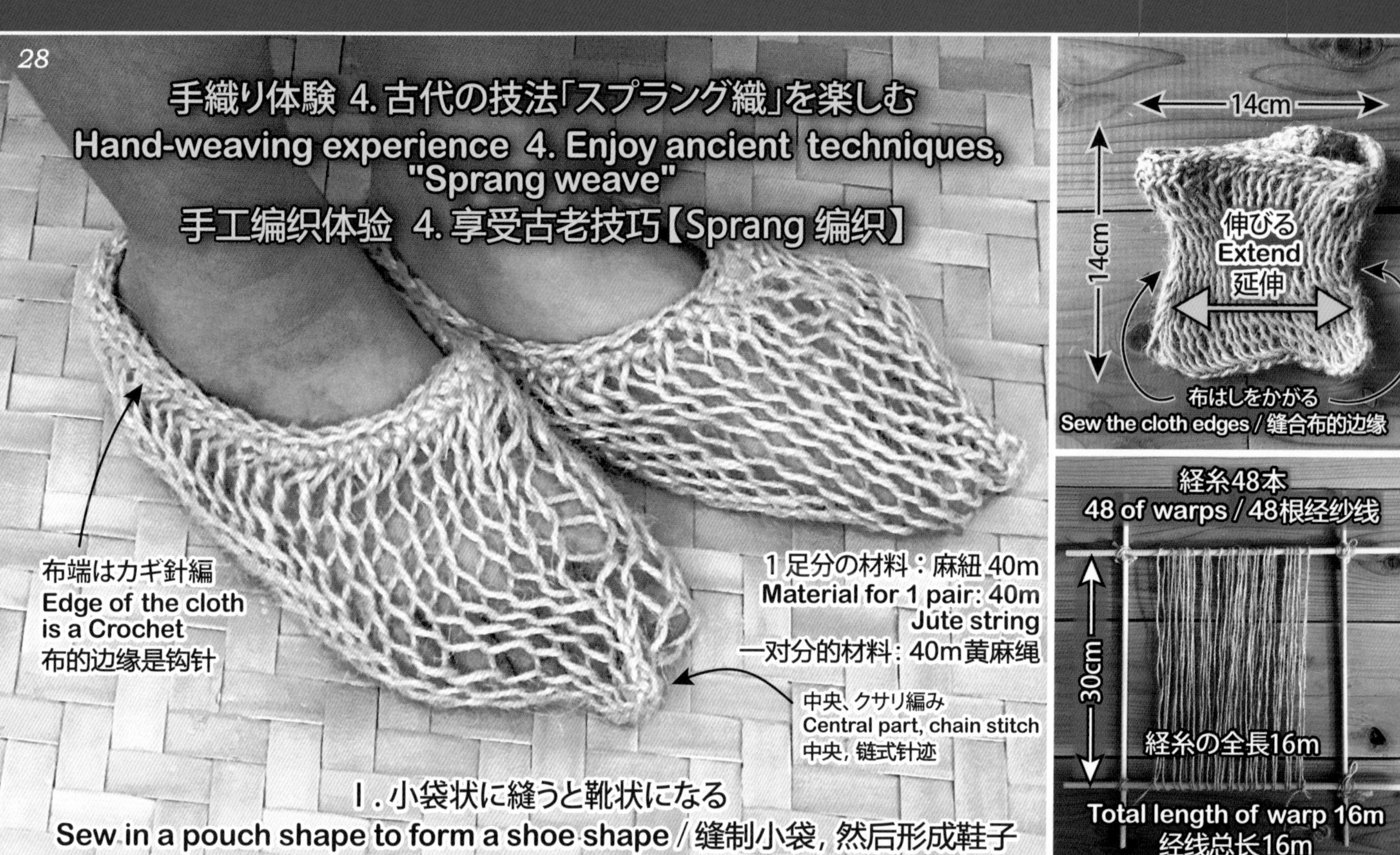

Ⅰ. 小袋状に縫うと靴状になる
Sew in a pouch shape to form a shoe shape / 缝制小袋，然后形成鞋子

Ⅱ. 筒状に縫うとレッグウオーマーになる
When sewn in a tubular shape, it becomes a leg warmer / 缝制管，会的在暖腿套

* サクサン糸：下村撚糸（京都）
http://y-shamoto.sakura.ne.jp/simo/s_index.htm
** キビソ糸（未精練）：WILD SILK MUSEUM

*Tussah silk thread: Shimomura Nenshi (Kyoto)
http://y-shamoto.sakura.ne.jp/simo/s_index.htm
**Kibiso silk yarn (undegummed): WILD SILK MUSEUM

* 柞蚕丝线：下村捻丝纱（京都）
http://y-shamoto.sakura.ne.jp/simo/s_index.htm
**Kibiso（条吐）真丝纱（非脱胶）：野生丝绸博物馆

III. 経糸に不向きな糸が楽しく織れる！
Can enjoy weaving threads that are not suitable for warp threads！
不适合经线的纱也可以织出有趣的东西！

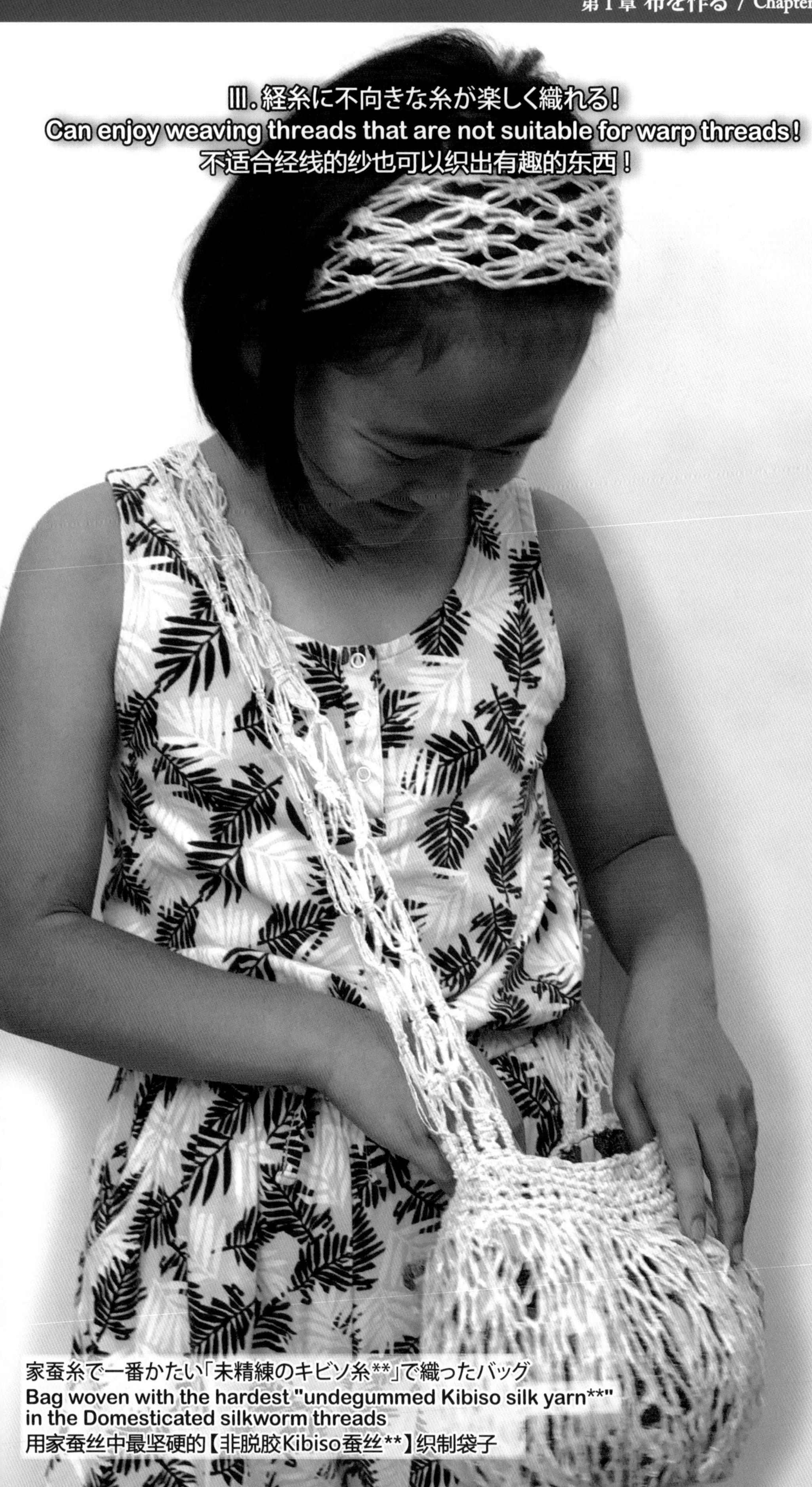

家蚕糸で一番かたい「未精練のキビソ糸**」で織ったバッグ
Bag woven with the hardest "undegummed Kibiso silk yarn**" in the Domesticated silkworm threads
用家蚕丝中最坚硬的【非脱胶Kibiso蚕丝**】织制袋子

太い細いが混在する絹糸で織った帽子
Hat woven with silk threads mixed with thick and thin
粗细丝线线混在纱制成的帽子

柔らかく、からみやすい糸で織った帽子
Hat woven with soft and easy to tangle silk threads
柔软而容易缠结丝线编织而成的帽子

上の帽子と同じ絹糸で織ったレッグウオーマー
Leg warmer woven with the same thread as the hat above
使用与帽子相同的丝线织成的腿部保暖套

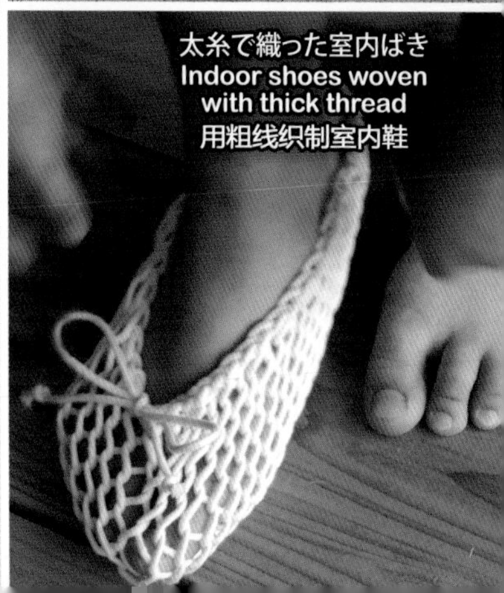

太糸で織った室内ばき
Indoor shoes woven with thick thread
用粗线织制室内鞋

Ⅳ. キビソのアートヤーンで帽子を作る
Making a hat with Kibiso silk art yarn / 用Kibiso真丝艺术纱制作帽子

1. ダンボール箱で織機を作る / Making a loom with a cardboard box / 用纸板箱制作织机

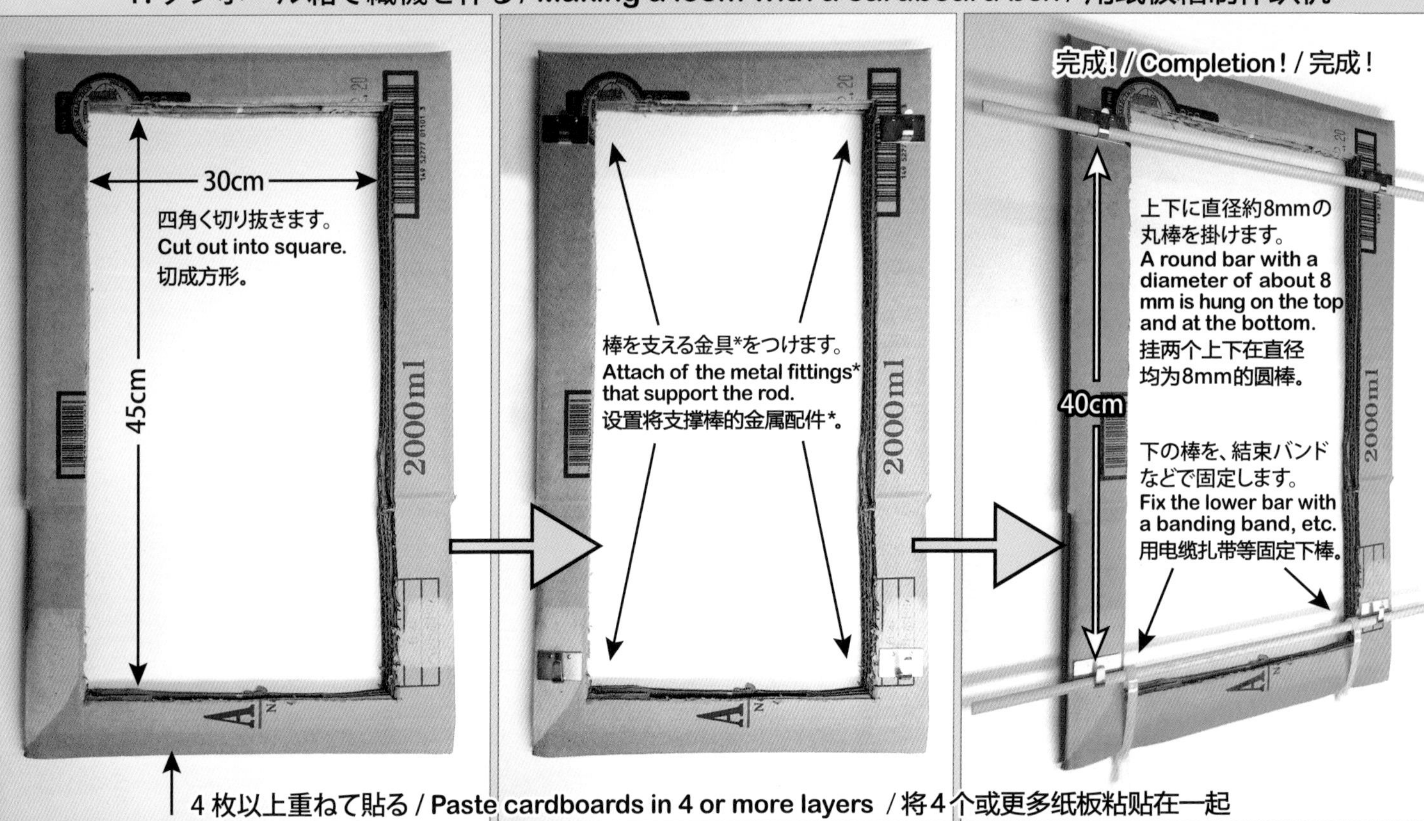

*棒を支える金具：ここでは電気配線工具、両面テープつきの電線支持用金具を使いました。結束バンドは、繰り返し使える物を選びましょう。

*Brackets for supporting the rods : Here we used the wire support bracket with double-sided tape that electric wiring tool. Choose a binding bundle that can be used repeatedly.

*支撑棒的金属支架：在这里，我们使用的是带有双面胶的用于支撑电线的金属支架。束缚带子可以选择可重复使用的东西。

2. 経糸を巻く / Wind the warp / 缠绕经线

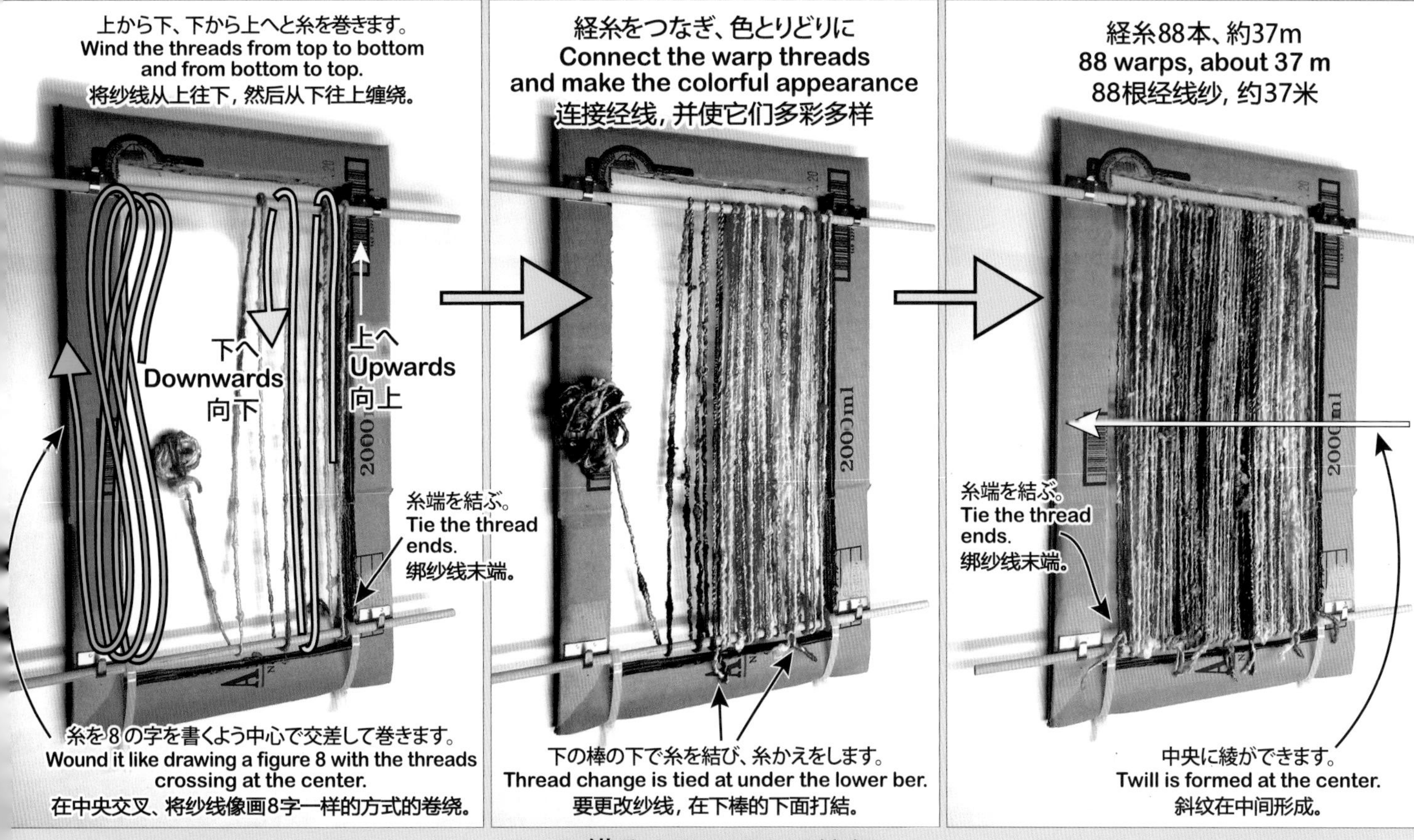

3. 織る / Weaving / 编织

経糸を上げ下げする道具として菜箸が便利
Bamboo cooking chopsticks are convenient as a tool to raise and lower the warp
竹制的烹饪筷子很方便用作升高和降低经线的工具

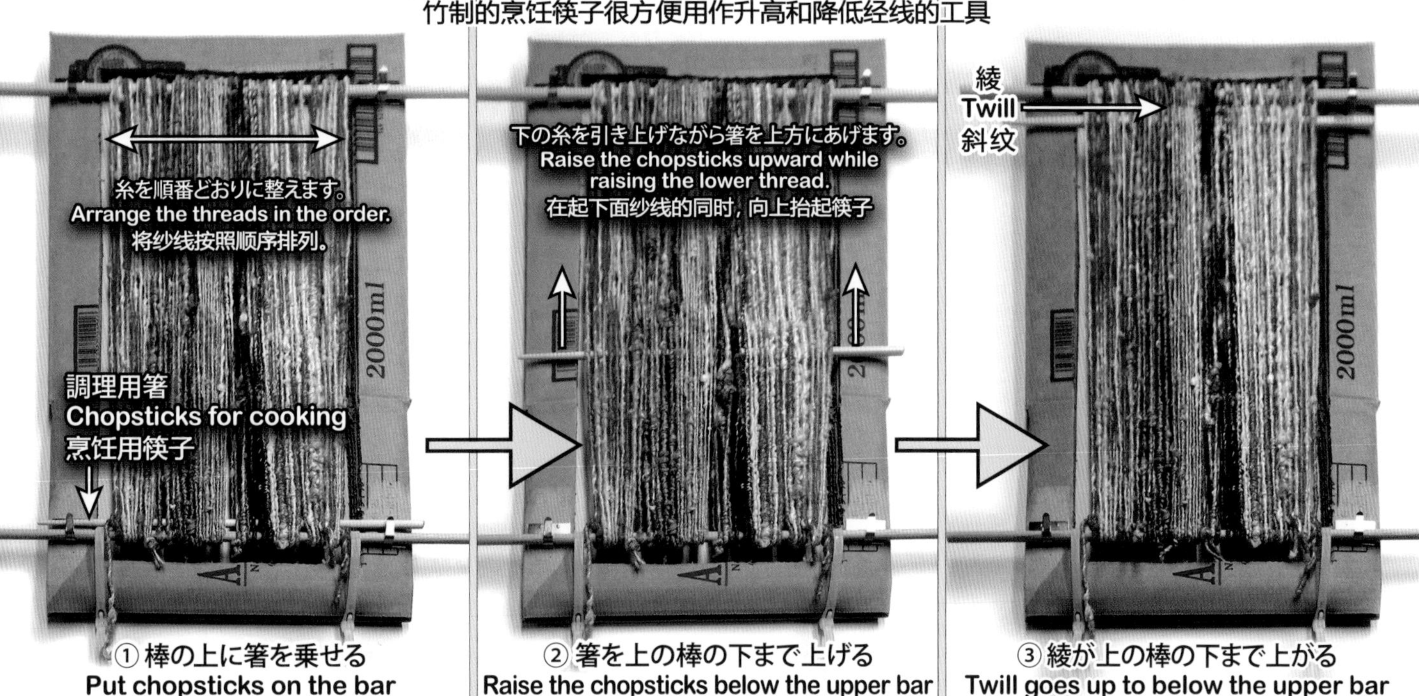

④経糸を交互に上げ下げし、糸をからめることで緯糸なしで布が織れる
It can be woven without weft by entangling warp while moving it alternately up and down
通过交替纠结上下的纱线，可以在不使用纬线的情况下进行编织

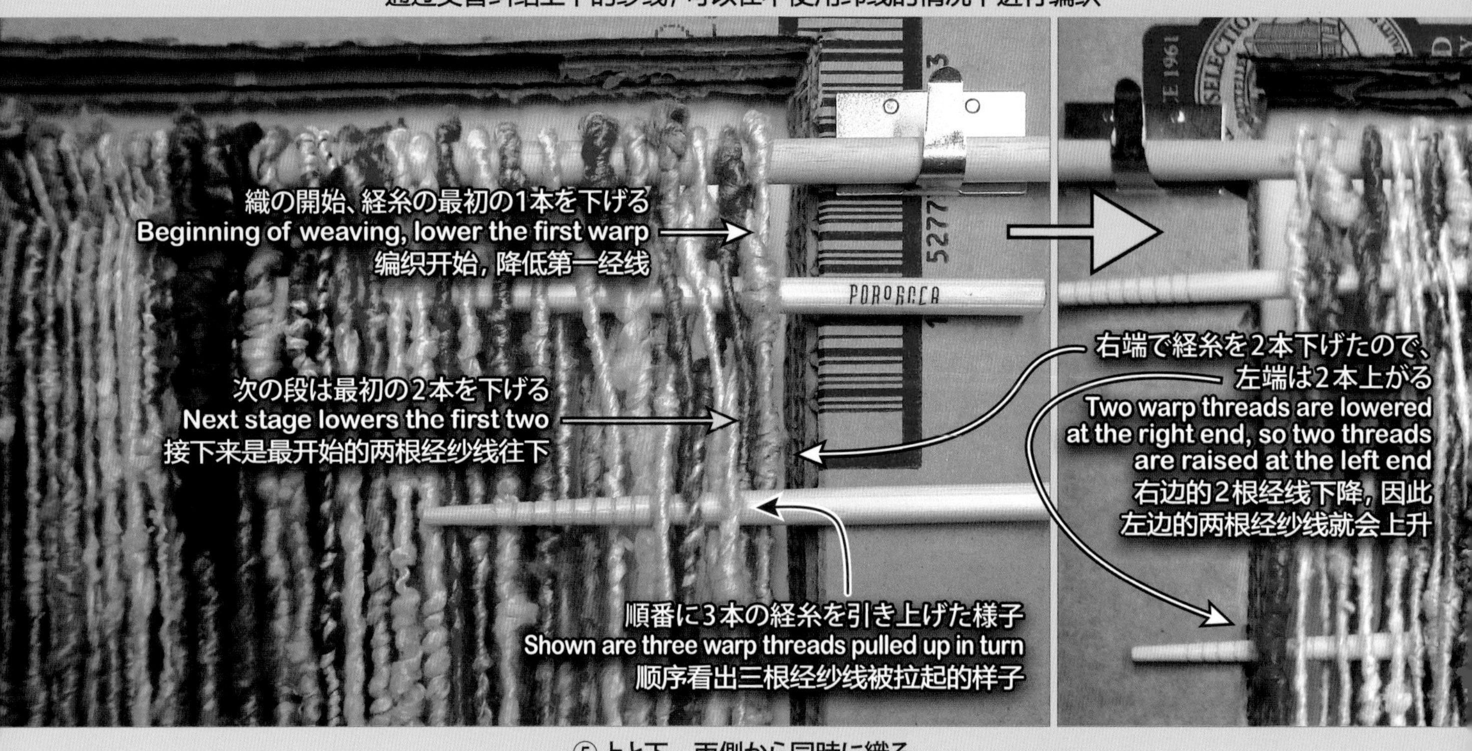

⑤上と下、両側から同時に織る
Weave from both top and botm simultaneously / 同时从在顶部和底部编织

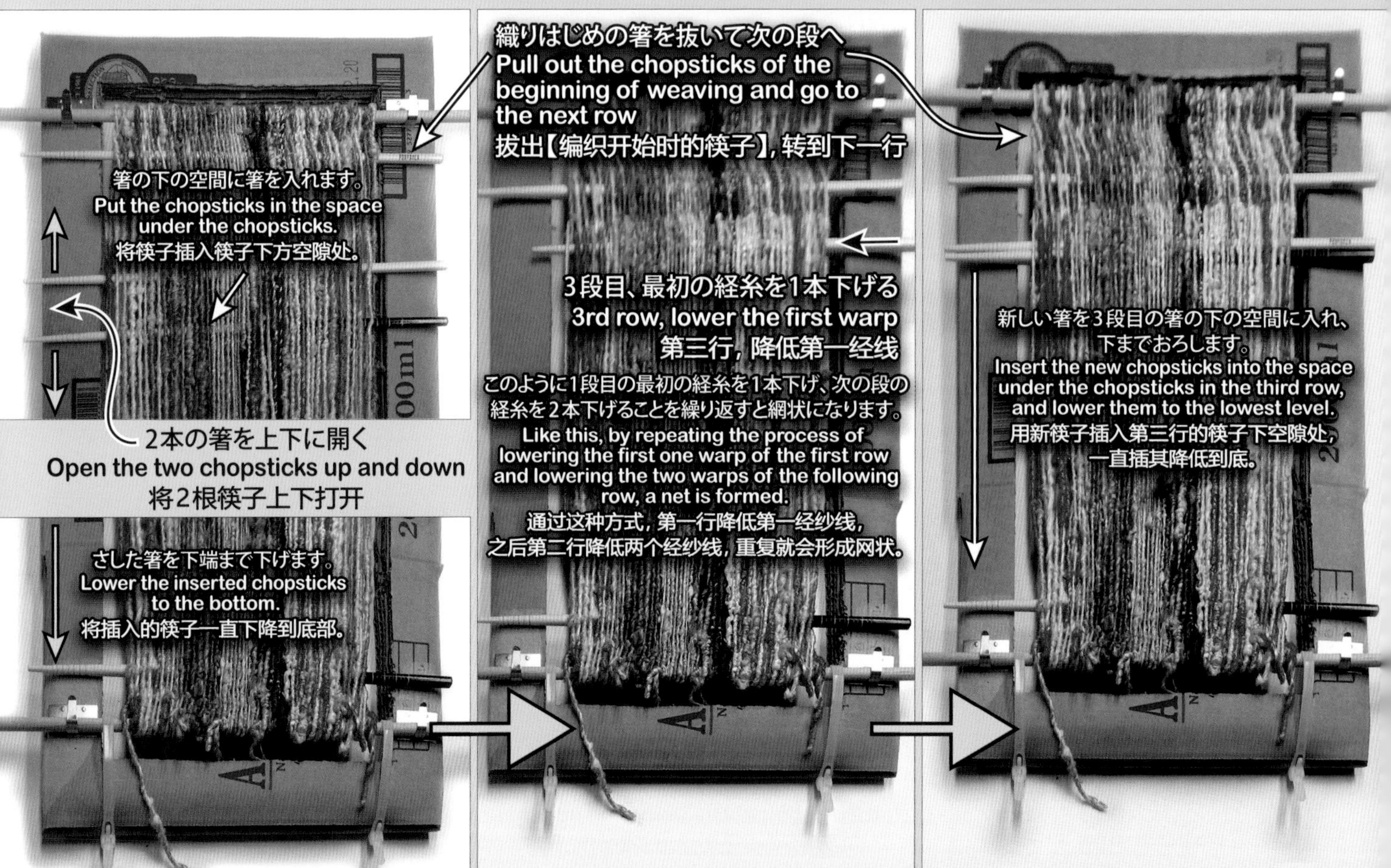

⑥ざっくりと魅力的な布が短時間で完成！
Attractive cloth is completed in a short time! / 只要很短时间，粗糙且却独特的布就完成！

4. 完成した布を織機からはずして帽子を作る
Remove the finished cloth from the loom to make a hat / 将织完的布从织机上取下，做成帽子

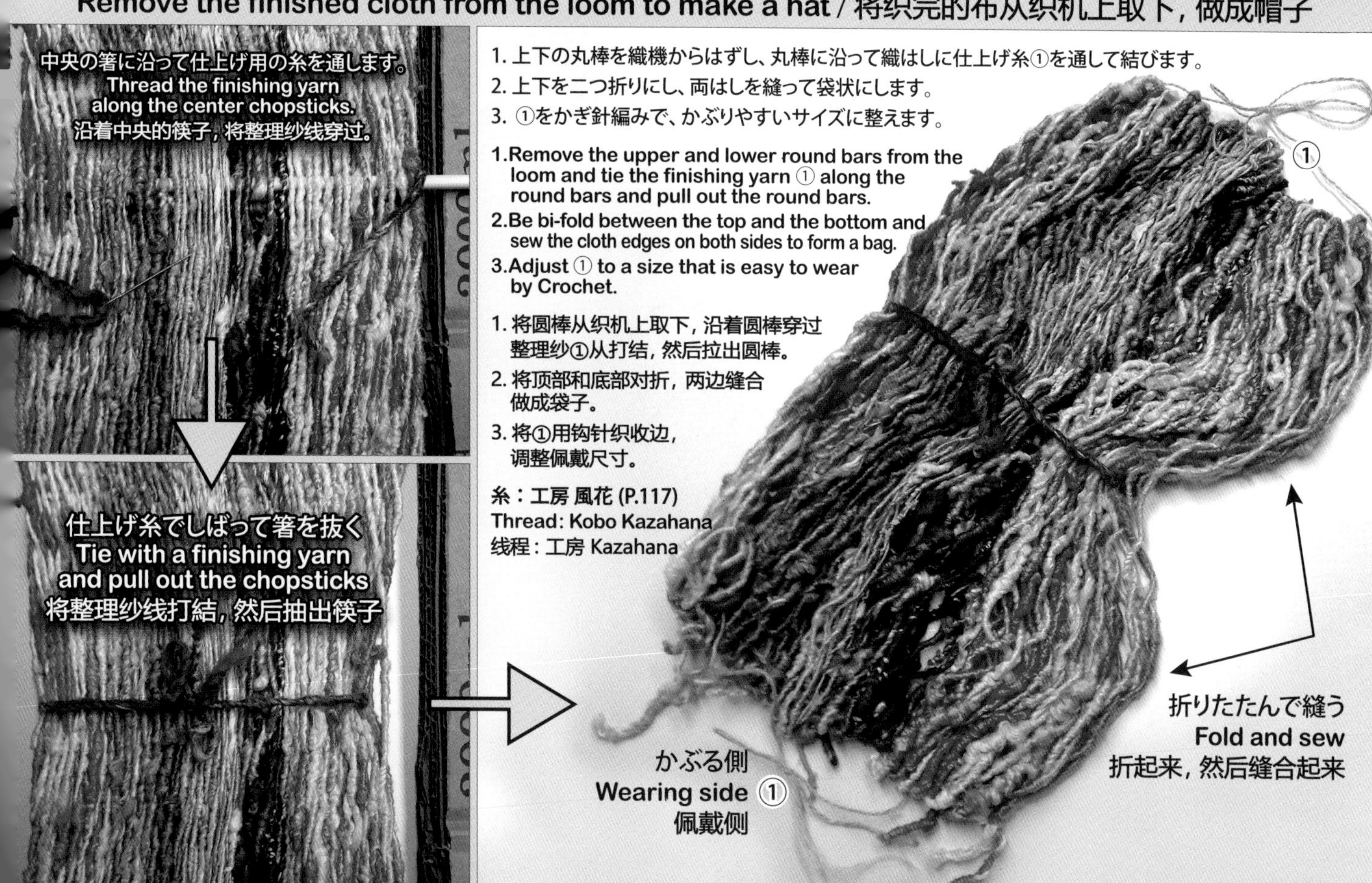

1. 上下の丸棒を織機からはずし、丸棒に沿って織はしに仕上げ糸①を通して結びます。
2. 上下を二つ折りにし、両はしを縫って袋状にします。
3. ①をかぎ針編みで、かぶりやすいサイズに整えます。

1. Remove the upper and lower round bars from the loom and tie the finishing yarn ① along the round bars and pull out the round bars.
2. Be bi-fold between the top and the bottom and sew the cloth edges on both sides to form a bag.
3. Adjust ① to a size that is easy to wear by Crochet.

1. 将圆棒从织机上取下，沿着圆棒穿过整理纱①从打结，然后拉出圆棒。
2. 将顶部和底部对折，两边缝合做成袋子。
3. 将①用钩针织收边，调整佩戴尺寸。

糸：工房 風花 (P.117)
Thread: Kobo Kazahana
线程：工房 Kazahana

①中央に通した仕上げ糸で経糸を3本ずつ束ねてチェーンステッチを編んで縮める
Use the central thread "finishing yarn" to bundle three warp threads at a time, and knit a chain stitch to shrink it
用中心的线【整理纱】, 一次捆扎三根经线, 并编织一个链式针迹以使其收缩

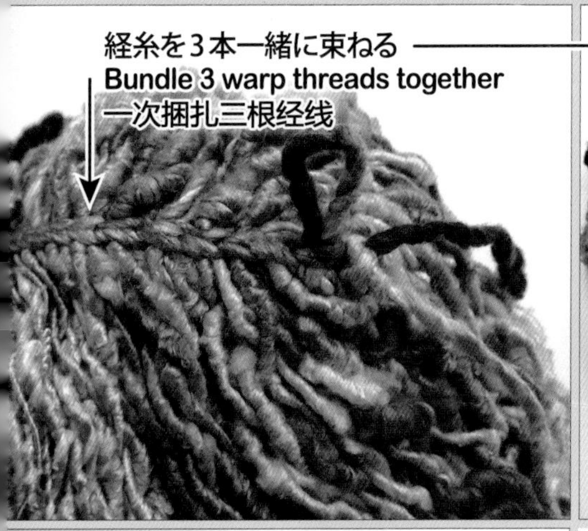

②飾りを作る
Make a decoration
做个装饰

道具を利用して
残り糸でポンポンを作る
Make pompom with remaining yarn using tools
利用工具, 将剩下的纱线做成绒球

裏面もかぶろう!
Let us wear the back side too!
让我们也穿内面!

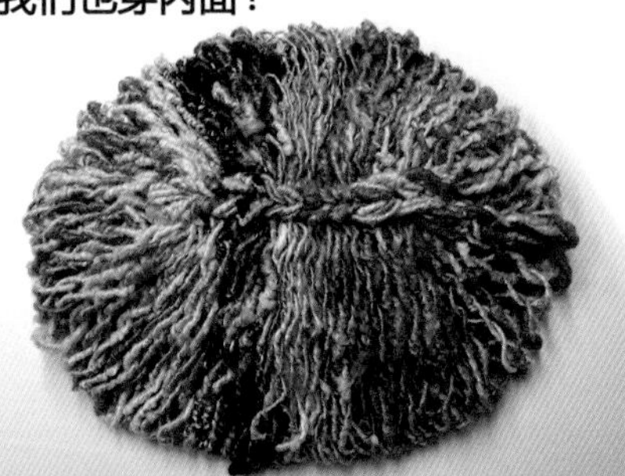

5. 完成! / **Completion!** / 完成!

V.キビソの未精錬糸でバッグを作る
Making a bag with Kibiso silk undegummed yarn / 用未脱胶Kibiso真丝线制作袋子

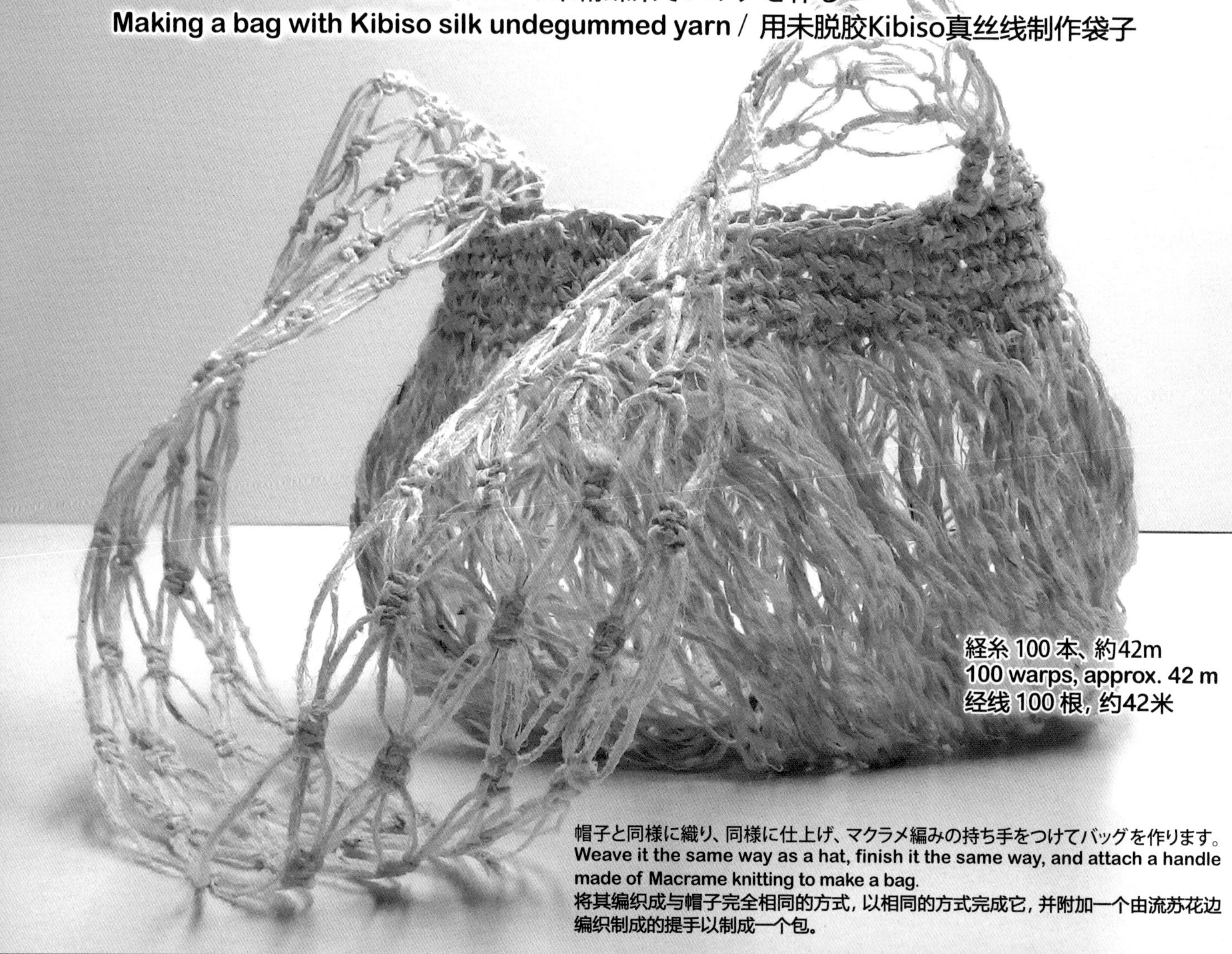

経糸 100 本、約42m
100 warps, approx. 42 m
经线 100 根，约42米

帽子と同様に織り、同様に仕上げ、マクラメ編みの持ち手をつけてバッグを作ります。
Weave it the same way as a hat, finish it the same way, and attach a handle made of Macrame knitting to make a bag.
将其编织成与帽子完全相同的方式，以相同的方式完成它，并附加一个由流苏花边编织制成的提手以制成一个包。

応用：縮緬の古布で中袋（下の写真）を縫い、持ち手を長くし、おそろいでヘアバンドも編みました（右の写真）。
Application: I sewed an inner bag (photo below) with old crepe cloth, lengthened the handle, and knitted a matching hair band (photo right).
应用：用旧的绉布缝制了一个内袋（如下图所示），加长了手柄，并以相同的方式编了发带（右图）。

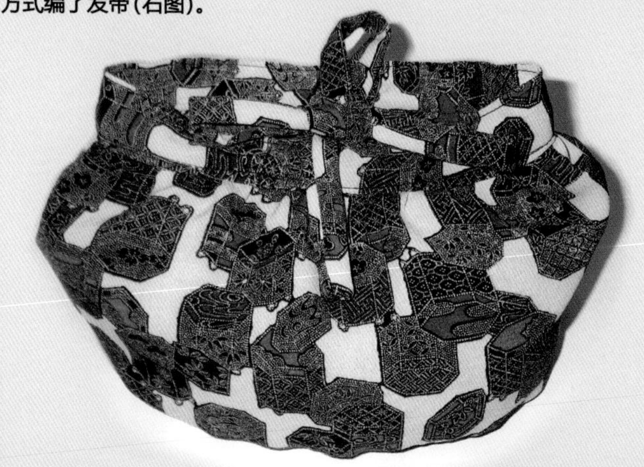

マクラメ編み 1.2m
Macrame knitting 1.2m
流苏花边编织编制 1.2 米

VI. 柔らかく、からみやすい「サクサン真綿糸」を織る / Weave "Tussuh floss silk thread" that is soft and easy to be entangle / 柔软但容易缠的【柞蚕丝绵线】织制

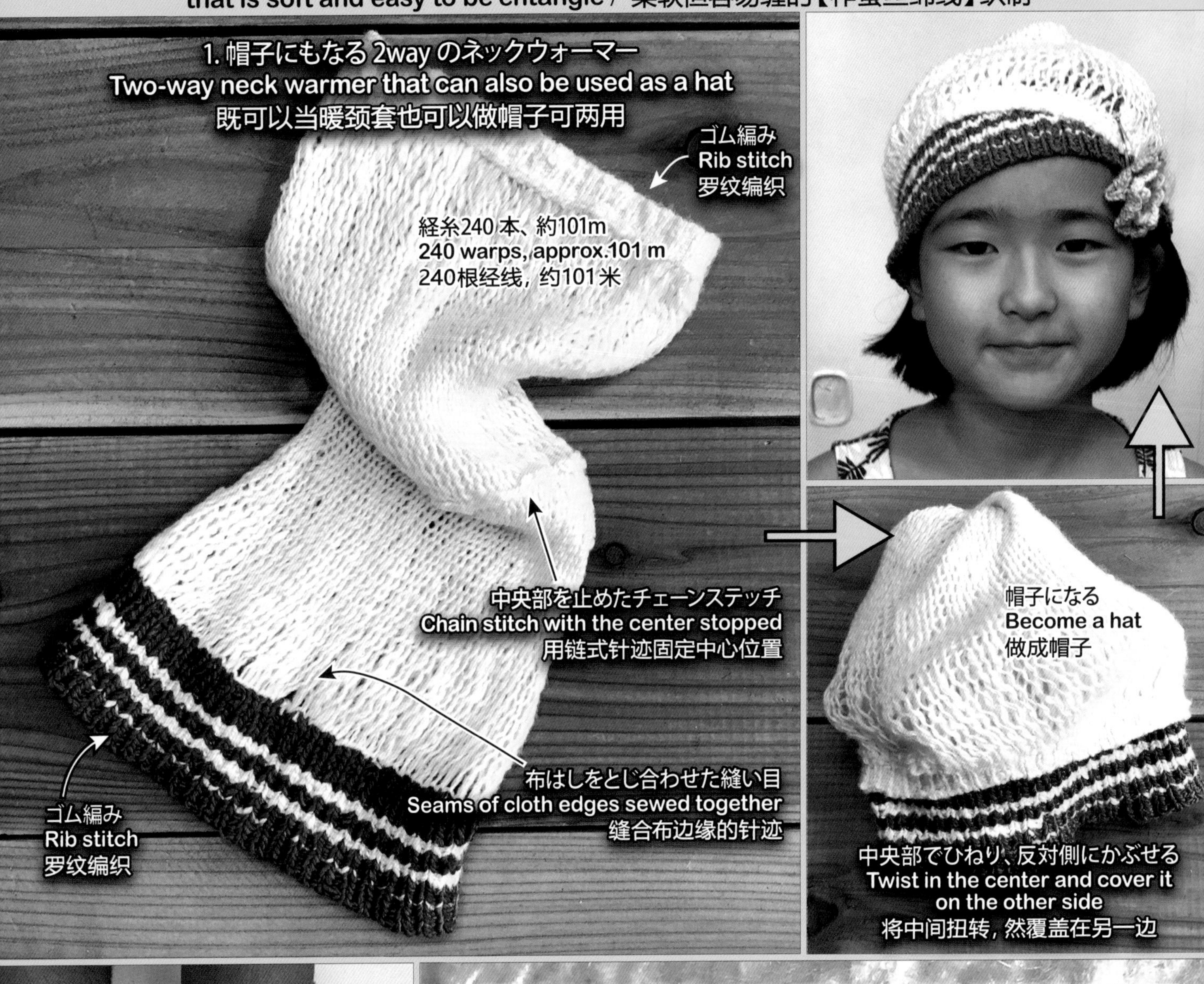

2. 真綿糸はからみやすいので、ていねいに糸を上下することが重要
As the floss silk thread is easy to be entangled, it is important to raise and lower it carefully
由于丝绵纱容易缠住缠结，因此小心地升高和降低螺纹很重要

3. からみやすい糸を密に織るときの注意
Precautions when densely weaving threads which is easy to be entangled / 在密织時要注意丝绵纱的缠绕

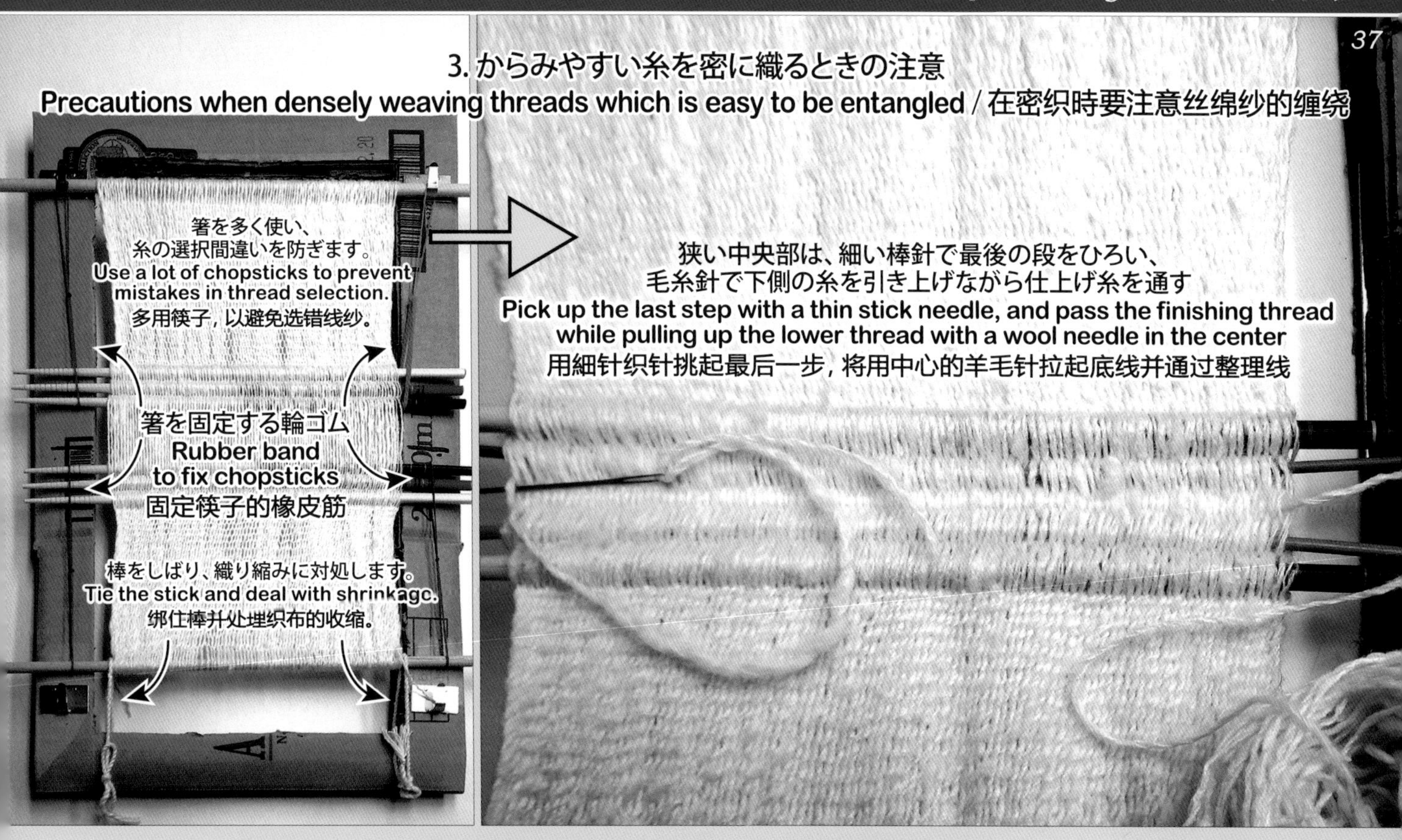

4. 伸縮性をさまたげずに中央部を固定する
Fixed the central part without hindering elasticity / 在不妨碍伸缩性的状况下固定中央部分

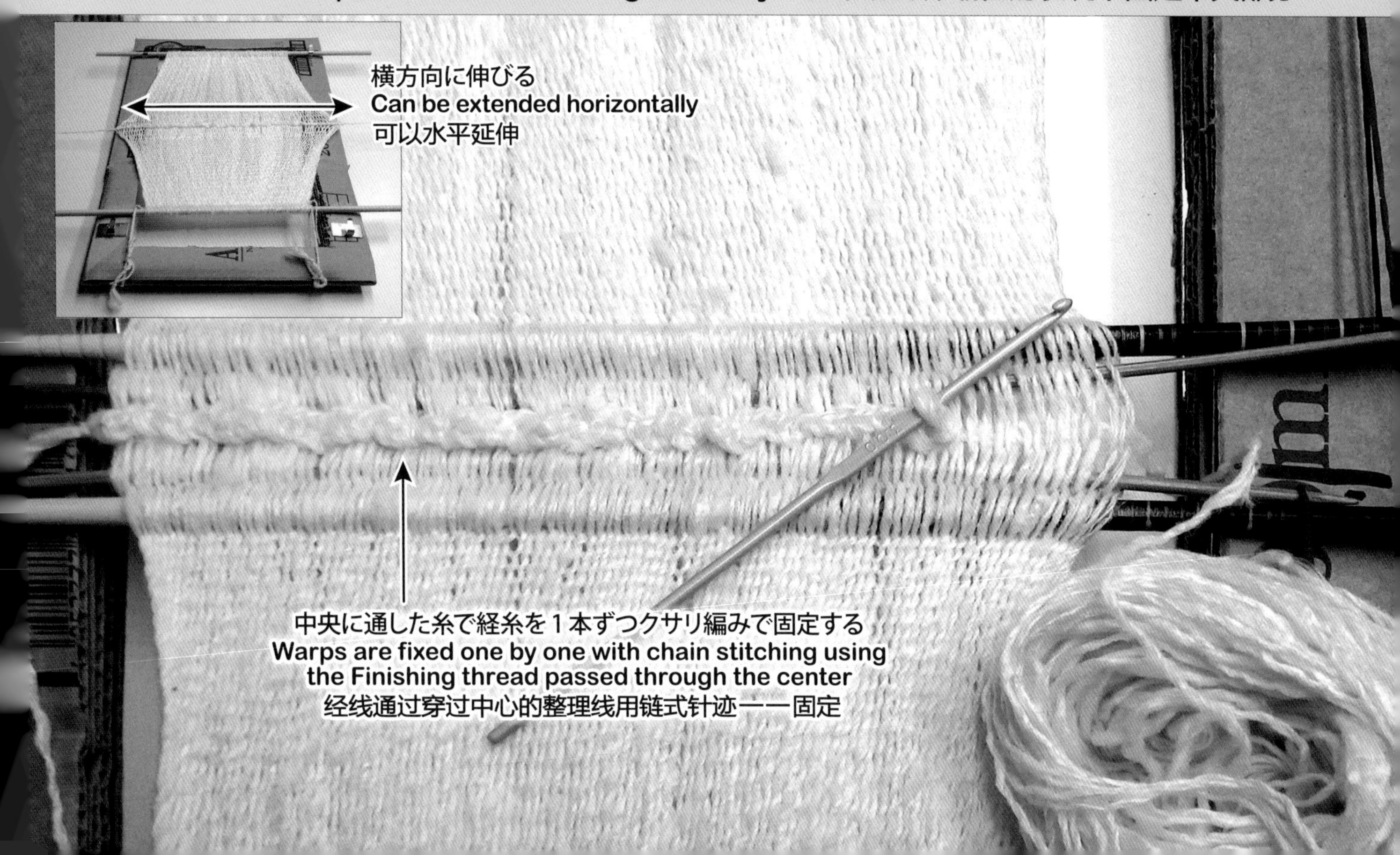

5. 丸棒から外し、布はしを仕上げる
Remove the cloth from the round bar and finish its edge / 从圆棒上取下，完成布的边端

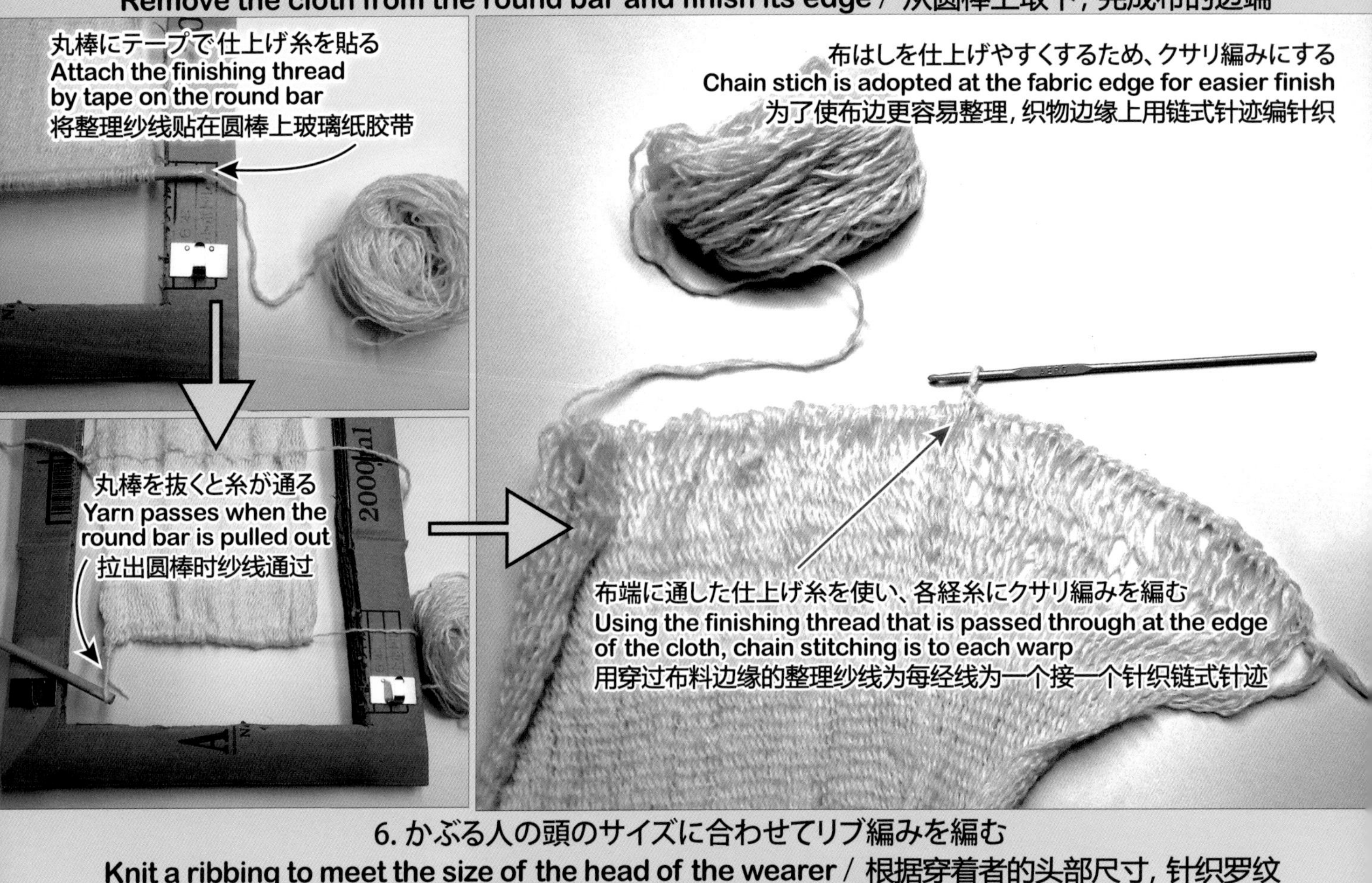

6. かぶる人の頭のサイズに合わせてリブ編みを編む
Knit a ribbing to meet the size of the head of the wearer / 根据穿着者的头部尺寸，针织罗纹

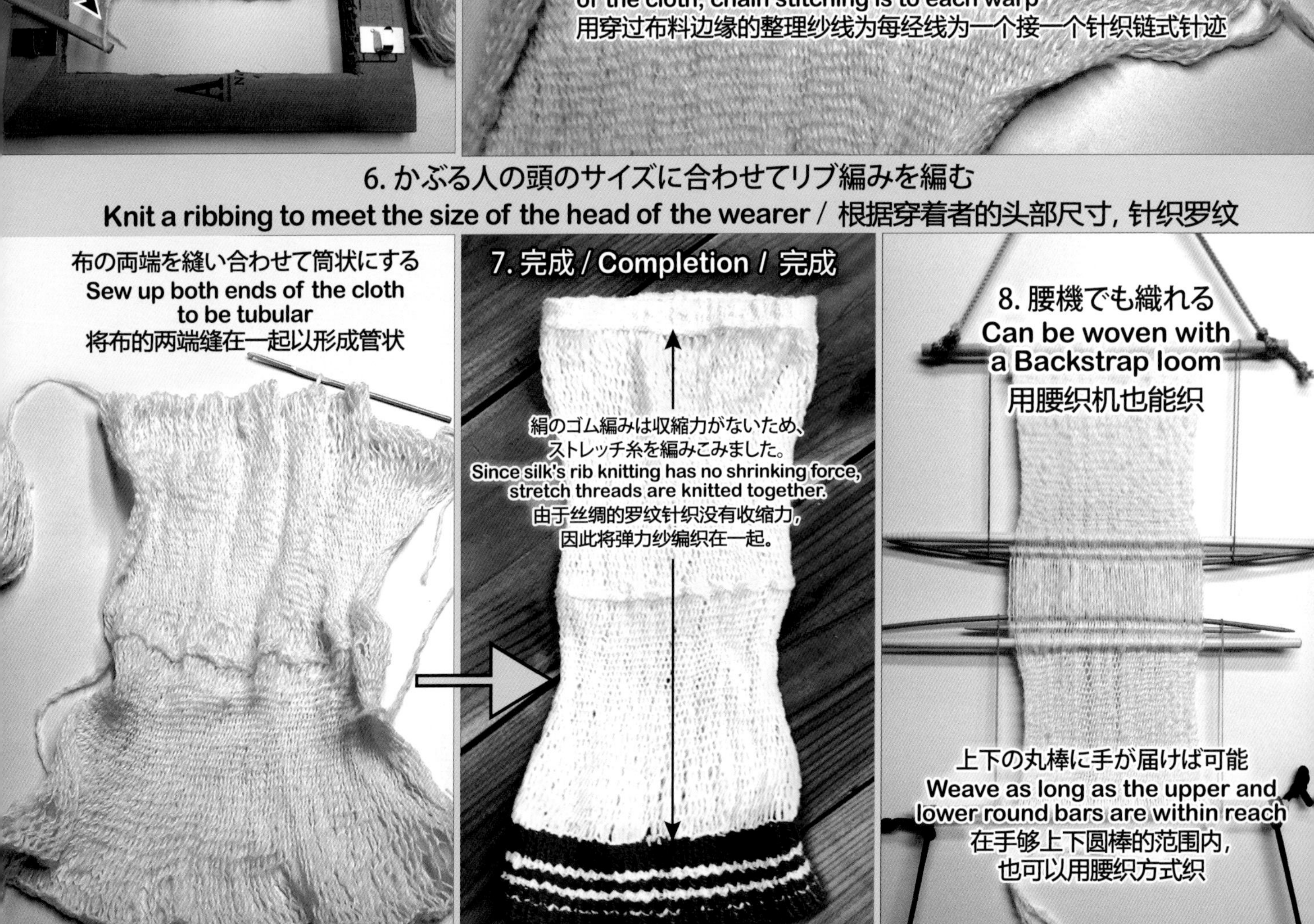

Ⅶ. 織りやすい綿コードで室内ばきを織る
Weaving indoor shoes with easy to weave cotton cord / 用容易织的棉绳线织室内鞋

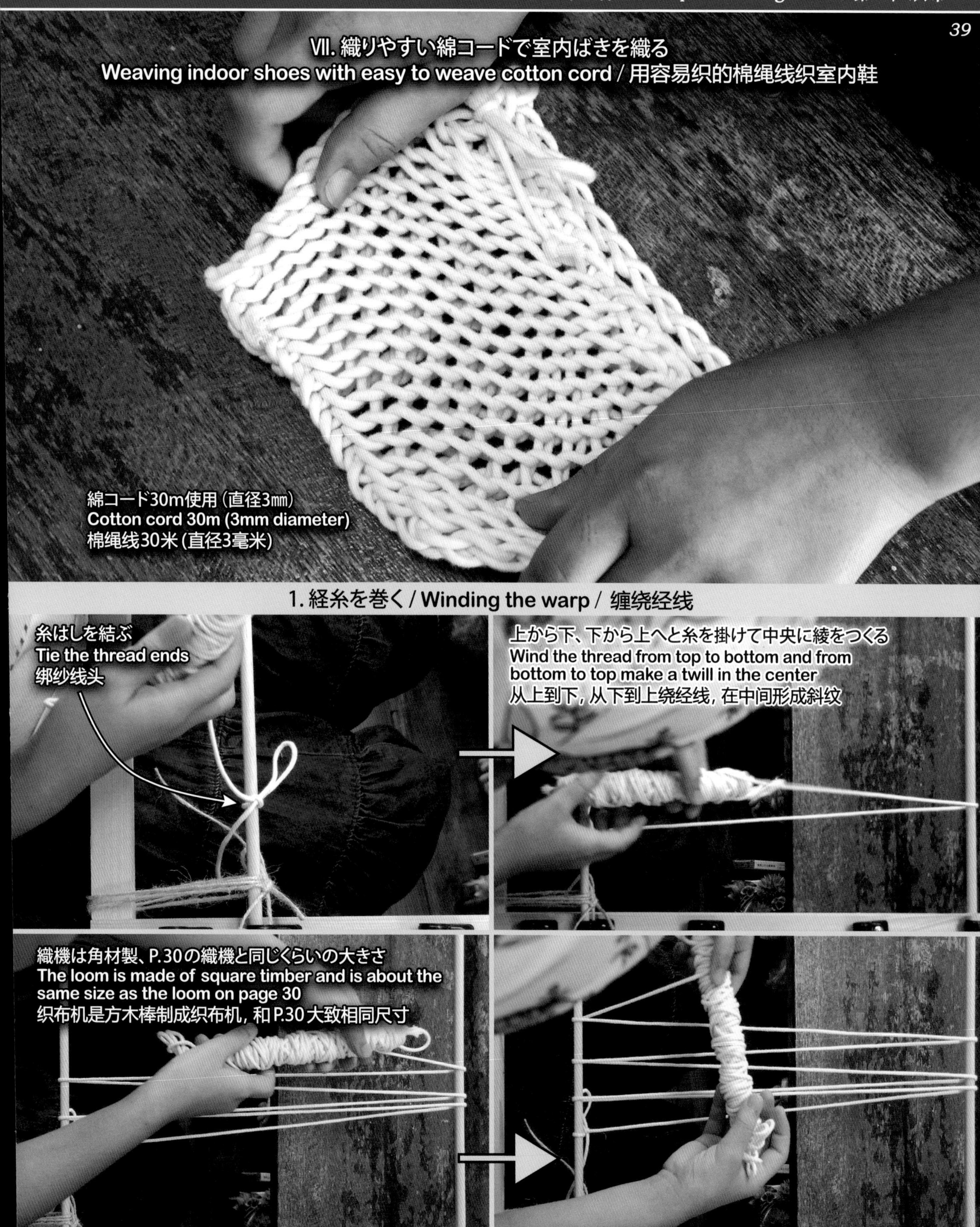

2. 規則正しく巻くことが重要 / Orderly winding of the warp is important / 井然有序的绕经线重要

経糸を19往復、38 本巻く
19 reciprocating windings threads, 38 warps
19 条往复式缠绕线紗, 38 条经线

3. 糸をそろえてから下の丸棒の上に箸を通す
After aligning the threads, pass the chopsticks above the round bar below
把紗线对齐后，在下面的圆棒上部穿过筷子

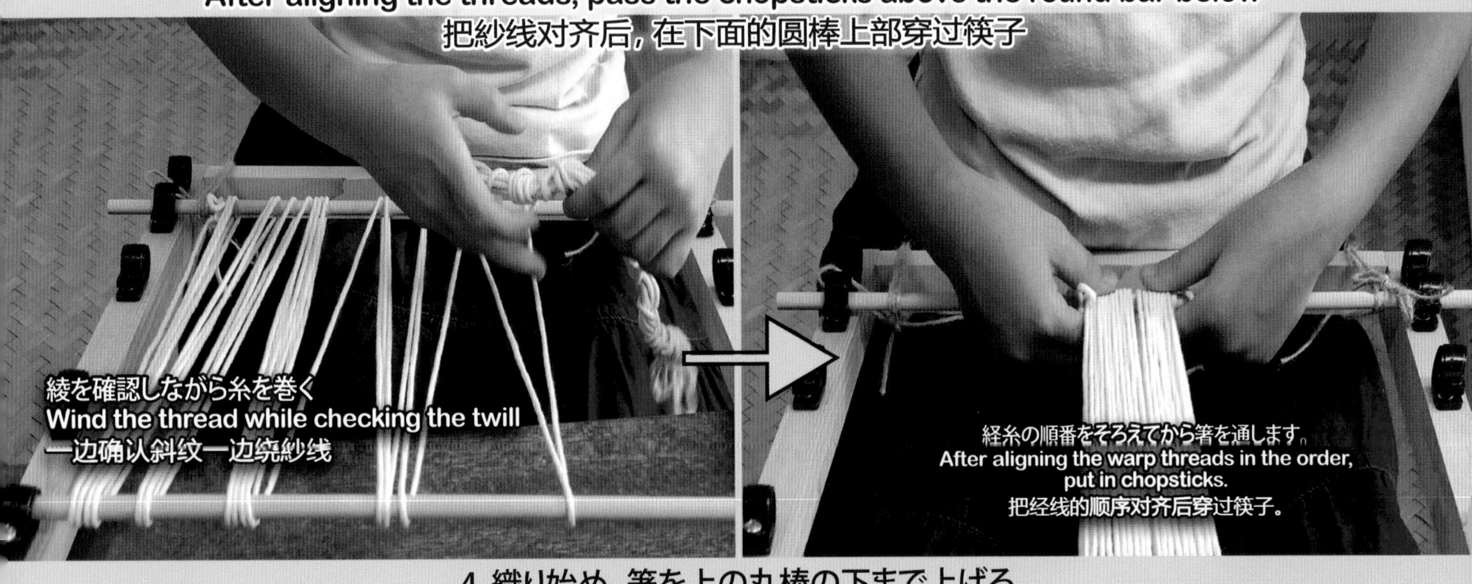

4. 織り始め、箸を上の丸棒の下まで上げる
Start of weaving, raise the chopsticks below the upper bar / 开始编织，把筷子提到上圆棒下面

*間違った例：織り始めに下げた糸と、織り終わりに上げた糸の本数が異なるときには、どこかで糸の選び方が間違っています。丁寧に見直して修正します。

*False example: If the numbers of threads lowered at the beginning of weaving and those raised at the end of weaving are different, the thread selection was wrong somewhere. Carefully review and correct.

*错误的示例：如果在编织开始时降低的线程数与编织结束时升高的线程数不同，那一定是在某处造错了线，仔细检查并更正。

5. 織り直し / Rewoven / 重织

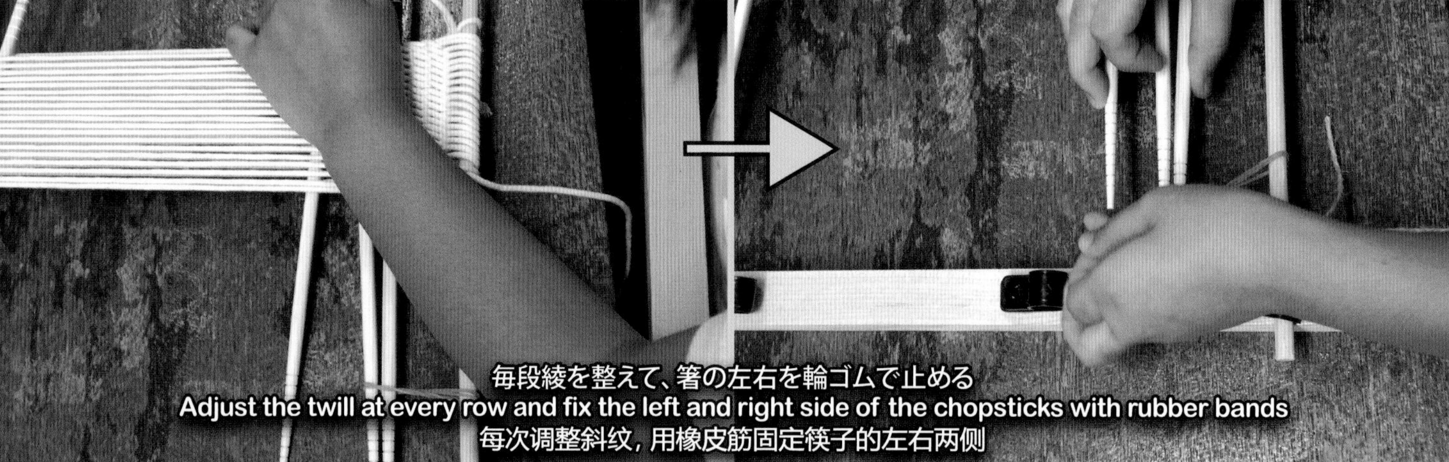

6. 中央部の織り方 / How to weave the central part / 中心部的编织法

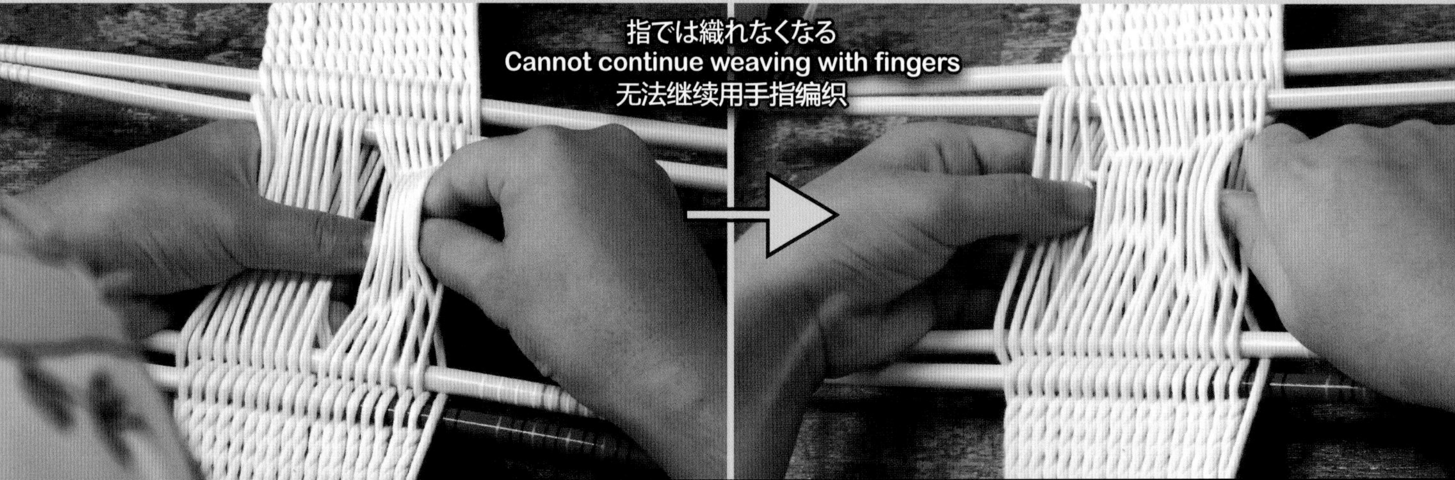

7. 中央部と両はしの仕上げ方
How to finish the center and the end of cloth / 中心部分和两端的完成方法

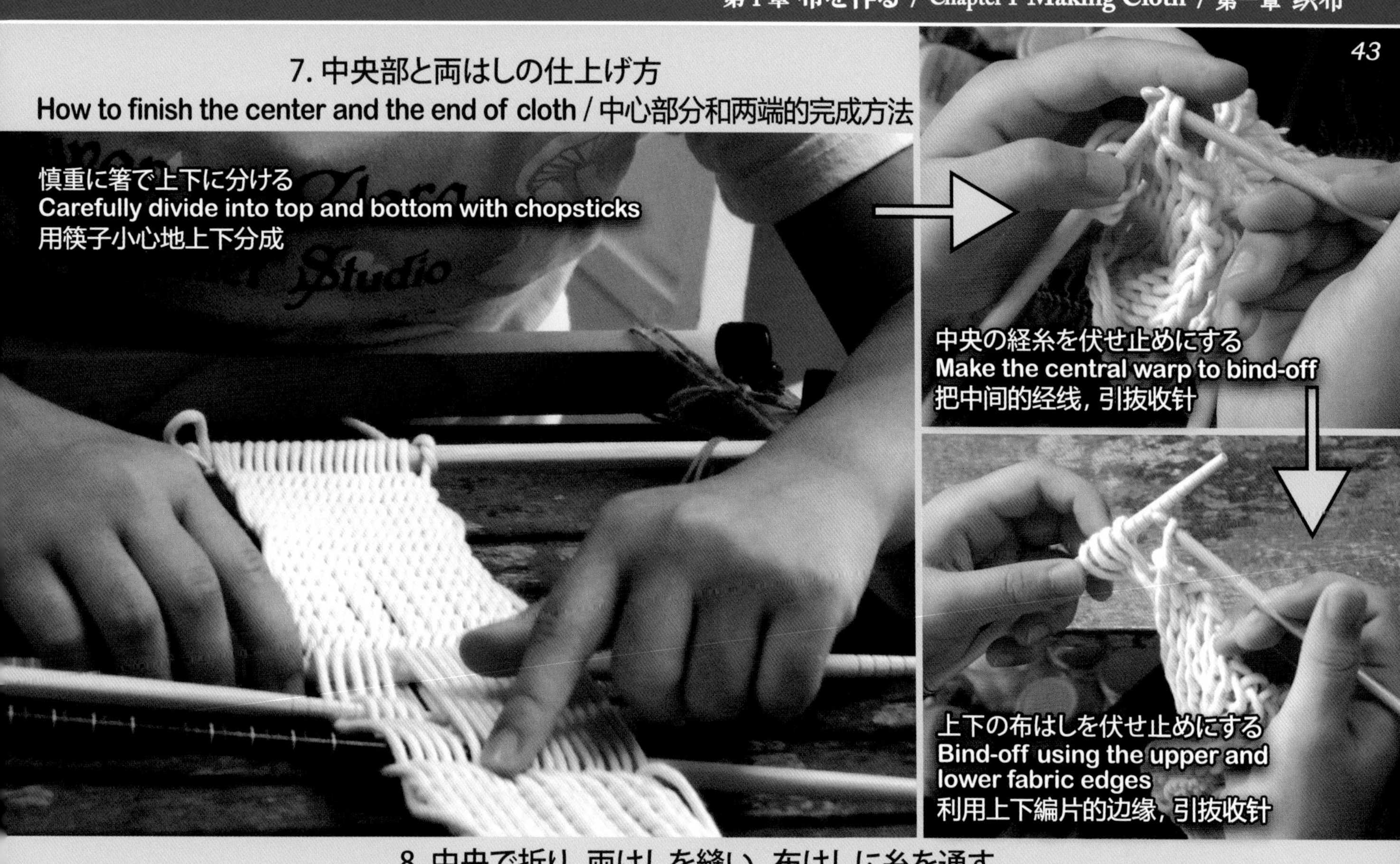

8. 中央で折り、両はしを縫い、布はしに糸を通す
Fold it in the center, sew both ends, and pass the Finishing thread of through the cloth edge
中間折起，两端缝上，然后在布边上穿过整理纱线

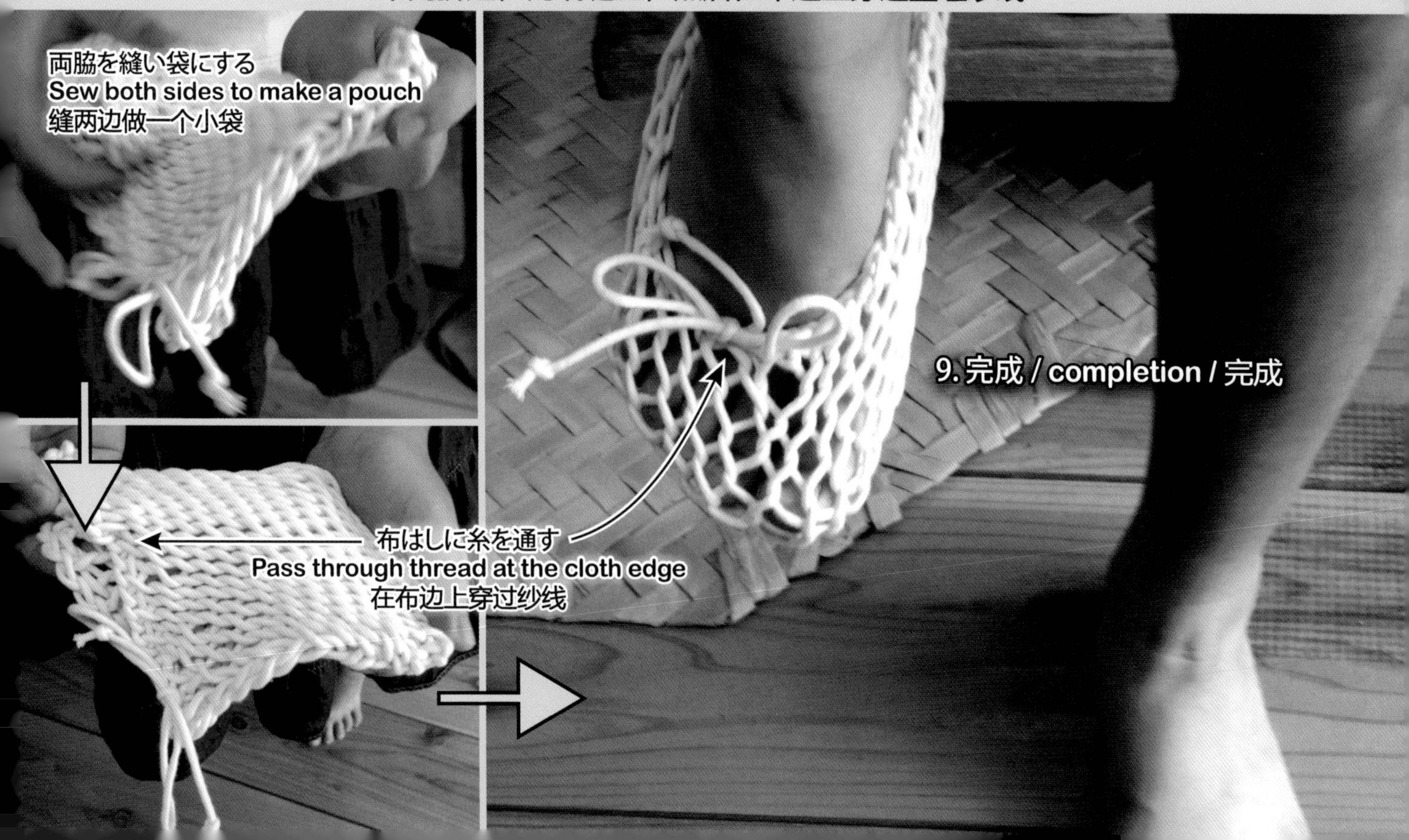

タンパク質繊維 / Protein fiber / 蛋白质纤维

絹はカイコの幼虫が自分の体を守る繭をつくるために、8の字を書くように首を振って吐いた糸です。その繊維は、人の毛髪やヒツジの毛糸と同じように動物性タンパク質でできているので、燃えると特有な臭いがします。

絹は火に近づけると、366℃まで発火しない燃えにくいタンパク質の繊維で、紙のように燃える木綿や、溶けながら燃える化学繊維とはまったく違います。

Silk is a thread which silkworm larvae spit out to form a cocoon to protect themselves while shaking their heads like a figure of eight. That fiber is made of animal protein, just like human hair and sheep's wool, so it has a unique smell when burned.

Silk is a protein fiber that is difficult to ignite up to 366° , even if put close to fire. It is completely different from cotton which burns like paper, and chemical fibers which burns while melting.

丝绸是蚕的幼虫为了制造能保护自己身体的茧、像写8字一样摇头吐出的线。它的纤维和人的头发和羊的毛线一样，都是由动物性蛋白质制成的，燃烧时会一种独特的気味。

丝绸是一种不易燃烧的蛋白质纤维，接近火时不会在366摄氏度以下点燃。与像纸一样燃烧的棉和一边熔化一边燃烧的化学纤维完全不同。

絹のタンパク質、セリシンとフィブロインの役割
Silk proteins, Roles of Sericin and Fibroin / 丝绸蛋白质，有丝胶与丝素的作用

カイコのアゴの左右に位置する一対の絹糸腺からフィブロイン(繊維質のタンパク質)が送り出され、中央に位置する吐糸管で糊状(にかわ質)タンパク質のセリシンに接着されて1本の糸になり、足場糸の毛羽や繭糸になります。

繭糸は、光沢や柔軟性のあるフィブロインをセリシンで固めてあるので、糸を繰って生糸にしてからセリシンをとり除きます。この工程を精練と呼びます。布を織った後、セリシンを落として布に風合をだす、羽二重やクレープなどの「後練り織物」、セリシンを落とした糸で仕上がりを見ながら織る「先練り織物」、セリシンを落とさずに織り、シャリ感や張りのあるまま使うオーガンジーなど、セリシンを利用して布に風合いを持たせます。

Fibroin (fibrous protein) is sent out from a pair of silk glands located on the left and right of the silkworm chin, and is attached to sericin, which is a pasty (glutinous) protein, at the central spinning tube to form a single thread. It becomes fluff or cocoon thread of scaffold thread.

Cocoon threads are made of lustrous and flexible fibroin and hardened with sericin, so threads are reeled to form raw silk and then sericin is removed. This process is called degumming. "Atoneri fabric" like Habutae (P.91) and Crepes (P.46) from which sericin is removed after weaving to give the cloth a texture; "Sakineri fabric" which is woven with threads from which sericin is removed beforehand and while checking the finished condition; Organdy, etc. are woven without removing sericin to give sharp and tight texture.

丝素蛋白(纤维蛋白)从位于蚕下巴左右的一对丝线腺中输送出来，在位于中央的吐丝管中与糊状蛋白的丝胶粘在一起形成单线，变成茧或脚手架线纱。因为茧丝是用丝胶固定有光泽和柔软的丝素，所以要把丝线缠绕成生丝，然后除去丝胶。这个工序叫做精练。编织后，除去丝胶蛋白，使面料具有质地的Habutae(纯白纺绸，电力纺，第91页)和Chirimen(绉纱，第46页)为呼称【精制过的丝织品，Atoneri织品】。用去除了丝胶的纱线机织的【Sakineri织品】。在不去除丝胶的情况下进行编织，可以增加面料的质感，就如带有爽滑度和张力的蝉翼纱(Organdy)。

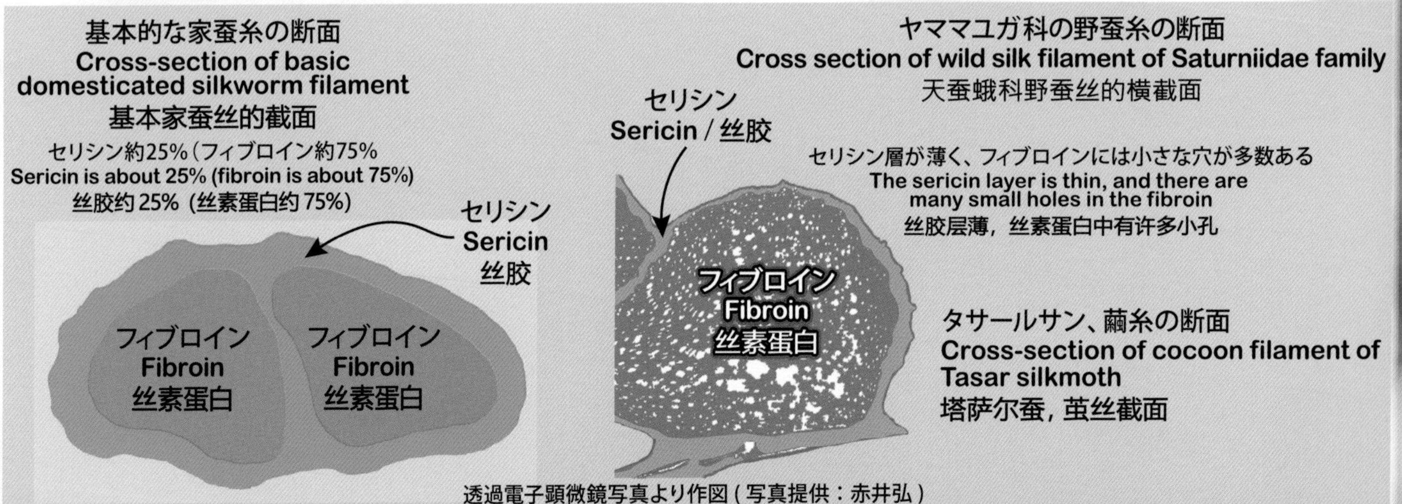

基本的な家蚕糸の断面
Cross-section of basic domesticated silkworm filament
基本家蚕丝的截面

セリシン約25%(フィブロイン約75%
Sericin is about 25% (fibroin is about 75%)
丝胶约25% (丝素蛋白约75%)

セリシン
Sericin
丝胶

フィブロイン
Fibroin
丝素蛋白

フィブロイン
Fibroin
丝素蛋白

ヤママユガ科の野蚕糸の断面
Cross section of wild silk filament of Saturniidae family
天蚕蛾科野蚕丝的横截面

セリシン
Sericin / 丝胶

セリシン層が薄く、フィブロインには小さな穴が多数ある
The sericin layer is thin, and there are many small holes in the fibroin
丝胶层薄，丝素蛋白中有许多小孔

フィブロイン
Fibroin
丝素蛋白

タサールサン、繭糸の断面
Cross-section of cocoon filament of Tasar silkmoth
塔萨尔蚕，茧丝截面

透過電子顕微鏡写真より作図(写真提供：赤井弘)
Drawing from transmission electron micrograph (Photo courtesy of Dr. AKAI Hiromu) / 来自透射电子显微照片画画(照片由AKAI Hiromu博士提供)

II. 絹糸の構造 / Structure of silk thread / 丝线纱的结构

カイコの繭1個の平均的な重さは2g、サナギの重さを引いた繭の重さは0.5gです。繭から引きだされる1本の糸の太さは1.5/1000mm、それは髪の毛の1/3程度の太さです。

21世紀初頭の合成糸の紡糸技術では、髪の毛の7500分の1ほどの細さのナノファイバーや、ストローのように内部が空洞の繊維で、通常の糸より多くの空気を含む中空糸もつくられていて、様々な質感をつくる技術があります。

しかし人は、2本のフィブロインを4層以上のセリシン層で包む、太さ0.015mm、長さ1000〜1500m* もの長繊維(フィラメント)を、化学合成できません。

The average weight of one silkworm cocoon is 2g, and the net weight of the cocoon after deducting the weight of pupa is 0.5g. Thickness of one thread pulled out from the cocoon is 1.5 / 1000 mm, which is about 1/3 of the thickness of hair.

At the beginning of the 21st century, there are various spinning technologies of synthetic yarns to create various kinds of texture, such as nanofibers that are about 1/ 7500 th of hair in fineness and hollow fibers that contain more air than ordinary yarns like straws.

However, humans cannot chemically synthesize long fibers (filaments) with a thickness of 0.015 mm and a length of 1000 to 1500 m * in which two fibroins are wrapped with four or more layers of sericin.

减去蛹的重量0.5克，一个也茧的本身平均重量为2克。从蚕茧中抽出的一根线的粗细是1.5毫米 /1000毫米，大约是头发的三分之一。

在在21世纪初的合成纤维纺丝技术中，有比头发细7500分之1左右的纳米纤维，和比普通的纤维含有更多空气的有像吸管一样内部中空的纤维等等，技术的发展使我们可以制造出各种各样质感的纺丝。

但是，人类无法化学合成厚0.015毫米，长1000至1500m*的长纤维(长丝)将两个丝心蛋白包裹着四层或更多层丝胶。

繭糸の模式図(セリシンは4層以上)
Schematic diagram of cocoon filament
(Sericin exists in 4 layers or more)
茧丝示意图 (丝胶为4层以上)

I
II
III
IV
繭
Cocoon
茧

繭糸1本の直径0.015mm
Diameter of one cocoon filament is 0.015 mm
一根茧丝的直径为 0.015 毫米

アミノ酸の基本配列
Basic sequence of amino acids
氨基酸的基本序列

シルクタンパク質はアミノ酸の単純な繰り返しで構成されています。
Silk protein consists of simple repeat of amino acids.
蚕丝蛋白是由氨基酸的简单重复组成的。

Gly	Ala	Gly	Ala	Gly	Ala	Gly	Ser	Tyr
グリシン	アラニン	グリシン	アラニン	グリシン	アラニン	グリシン	セリン	チロシン
Glycine	Alanine	Glycine	Alanine	Glycine	Alanine	Glycine	Serine	Tyrosine
甘氨酸	丙氨酸	甘氨酸	丙氨酸	甘氨酸	丙氨酸	甘氨酸	丝氨酸	酪氨酸

基本配列のアミノ酸は結晶領域(疎水性が多い)を結成し、非晶領域のアミノ酸(親水性)が基本配列のアミノ酸をつなぐ。
Crystalline regions (mostly hydrophobic) and amorphous regions (hydrophilic) exist. The former consists of basic amino acid sequence and amino acids in the latter connect those in the former.
基本序列的氨基酸组成结晶区域 (主要是疏水性的)，非晶区域的氨基酸(亲水性)连接基本序列的氨基酸。

繭から糸を巻きあげると生糸、生糸を精練するとシルクになる
Thread when reeled up from the cocoon is raw silk, which becomes silk after being degummed.
由蚕茧卷线是生丝，然后对生丝进行脱胶，则变成了丝

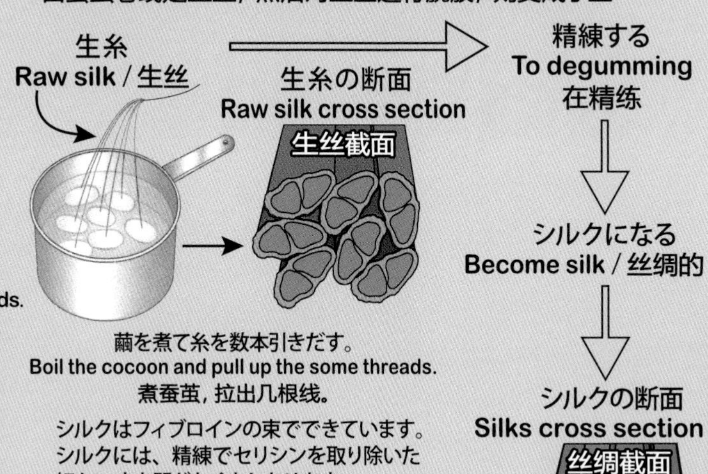

丝绸截面

シルクはフィブロインの束でできています。シルクには、精練でセリシンを取り除いた細かいすき間がたくさんあります。
Silk is made of a bunch of fibroins. Silk has a lot of fine gaps between silk made by sericin removal.
丝绸是由丝素束组成的。精练除去丝胶因蛋白的丝有许多细微的间隙。

シルクプロテインに結晶領域と非晶領域があることから、シルクの強度と弾性が生まれ、疎水性と親水性があることでシルクの特性が現れます。

結晶領域の分子は強く引き合い、凝縮しようとするために引っ張る力が鋼鉄よりも強いといわれ、弾力性に富んでいます。

Silk protein has crystalline and amorphous regions, which give silk its strength and elasticity. Existence of both hydrophobic and hydrophilic regions gives silk its characteristics. Since molecules in the crystalline region strongly attract each other and are apt to clump, it is said that they have a stronger pulling force than steel and are highly elastic.

蚕丝蛋白具有晶体区和非晶区，产生了蚕丝的强度和弹性，疏水性和亲水性表现出蚕丝的特性。晶体区域的分子被强烈吸引，据说为了凝缩而产生的拉力比钢铁强，因此富有弹性。

＊1000〜1500m：カイコの品種は、糸が多くとれるように改良されてきました。しかし、特殊な糸の場合は短く、300mくらいの品種もあります。

＊1000-1500m: Silkworm has been breed improvement so that many threads can be taken. However, in the case of special varieties, there is also a yarn of about 300m.

＊1000-1500m：蚕的品种经过改良，可以收获更多的线。但是特殊品种，有些长度约300m。

非晶領域では分子の並び方が不規則です。伸縮性があって柔らかく、空気を含み、不規則な分子の間に水分を多く吸収し、絹を着てしばらくすると体が温まる保温効果を発揮します。また、一定の湿度を吸収すると放湿をはじめる、溜池的な機能と、放湿時に発生する気化熱が、上がり過ぎた温度をさます役割を果たしています。

絹が持つこれらの特性が、絹を着るとほのかに暖かく、蒸れずにべとつかないという気心地を作り出しています。また、 絹を構成するアミノ酸分子の幾種類もの親水基は、絹の優れた染色性に役立ちます。

In the amorphous region of fibroin, the arrangement of molecules is irregular. Silk is elastic and soft, contains air, absorbs a lot of water among irregular molecules, makes body warm soon after wearing silk clothes and has an effect of keeping body temperature. In addition, when it absorbs a certain amount of humidity, it starts releasing moisture, that is, it functions like a reservoir on the one hand, and on the other hand, when moisture is released, it robs heat by vaporization, which leads to the control of body temperature that has risen too much. These properties of silk create a feeling that when people wear silk fabric, they have feeling of being comfortably warm, not damp and not sticky. Various hydrophilic groups of the amino acid molecules which consist fibroin protein also contribute to its excellent stainability.

在非晶区，分子排列是不规则的。它具有伸缩性，柔软，透气，在不规则的分子之间吸收水分较多，一穿上蚕丝身体就感到柔和感，表现出保温效果。另外，吸收一定的湿度后开始放湿，具有贮水池的功能，放湿时产生的气化热，也起着控制温度升高的作用。

丝绸所具有的这些特性，使得穿上丝绸时会产生出些许温暖、不闷不粘的感觉。此外，构成丝绸的氨基酸分子中的几种亲水性基团有助于丝绸的优良染色性能。

縮緬の表面の枝毛と毛玉
Branch fiber and fiber ball on surface of crepe
绉绸表面的分支纤维（枝毛）和纤维球

原寸
Actual size
实际尺寸

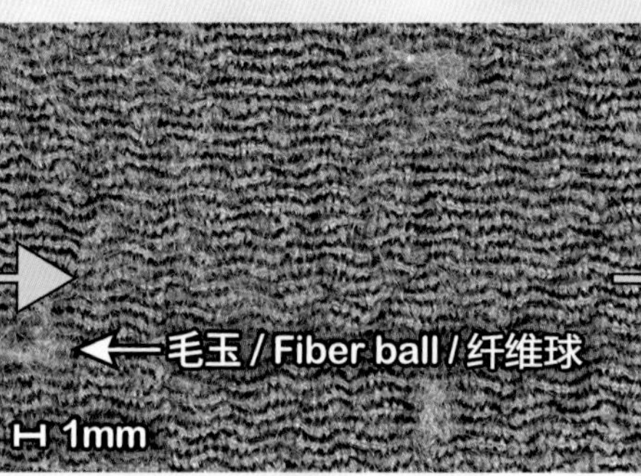

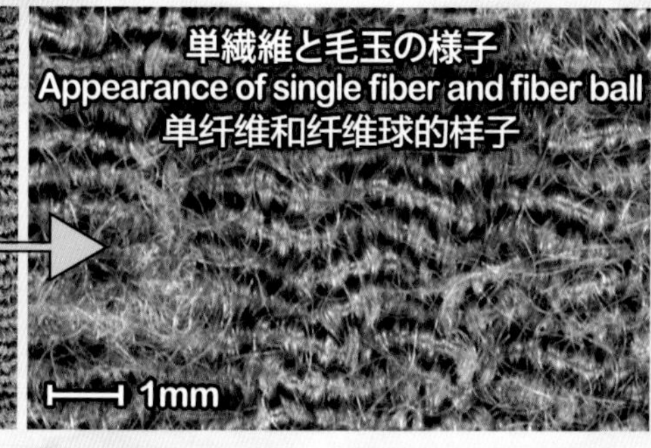

シルクに繰り返し摩擦が加わると、単繊維（フィブロイン）が引き出されて枝毛として現れます。更に摩擦が繰り返されると、フィブロインがちぎれて絡み合い毛玉になります。この状態が繰り返されると、フィブロインがちぎれて布が劣化し、穴や破れが生じます。

絹が毛玉になって、布から糸が強く引かれていたらなるべく戻し、それ以上糸を引かないように注意しながら毛玉を切ります。気をつけて手入れをすれば、長く着ることができます。

また洗濯でのもみ洗いは布が劣化するので、洗液に浸して軽く押し洗いをしましょう。

When friction is repeatedly applied to silk, single fibers (fibroin) are pulled out and appear as branched fiber. When the friction is further repeated, fibroin is torn and entangled to form a fiber ball. If this is repeated, cloth will deteriorate, resulting in holes and tears. If the silk becomes a fiber ball and the thread is pulled strongly from the cloth, return it as much as possible, and cut the fiber ball while being careful not to pull the thread any more.

Also, since silk fabric deteriorates when rubbed by hand washing, soak it in the washing liquid and press it lightly.

如果丝绸反复摩擦，单纤维（丝素）就会被拉出，出现枝毛分乱叉线纱端。再反复摩擦的话，就会被撕下缠绕在一起，形成纤维球。如果反复这样做，布就会劣化，形成洞和撕裂。

如果丝绸变成了纤维球，而且从面料上被拉得很紧的话，尽量把它拉回来，一边注意不要再拉线一边剪纤维球。小心打理的话，可以穿很长时间。另外，洗涤时如果揉搓洗涤的，会劣化，所以请浸泡在清洗液中轻轻按压洗涤。

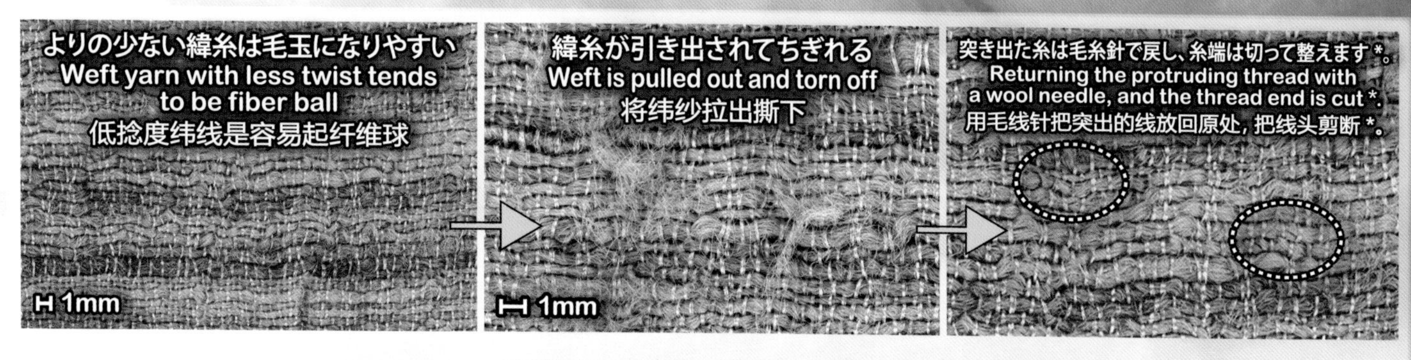

III. 繭から絹糸を作る5つの方法
Five ways to make silk thread from cocoons / 用茧制成丝线的五种方法

古代の人々は野にある様々な繭を集めて、知恵と工夫で糸にしてきました。

5,000年くらい前から養蚕に成功し、大量の糸を安定して作ることが可能となり、明治時代には、富岡製糸場に代表されるような工場で生糸が生産されるようになりました。

家畜化されていない各種野蚕の繭は、種類によっては生糸が繰れず、紬糸(紡糸)にしかならない繭もあり、今でも熟練した人が1粒ずつ糸を繰っています。

それでは繭からどのようにして糸を繰り、どのような種類の糸ができるのでしょうか?

それらのおおよそについて、着る側からの観点に立って紹介しましょう。

Ancient people have collected various cocoons in nature and have made them into threads using their wisdom and ingenuity.

Around 5,000 years ago, domestication of silkworm was successful, and it became possible to stably collect a large amount of thread. In the Meiji era, Raw silk came to be produced in factories as represented by Tomioka Silk Mill.

Among cocoons of various wild silkworms that have not been domesticated, some cannot be made into raw silk depending on species, and some can only be unevenly spun silk thread (spinning). Even today, skilled people spin thread out of cocoons one by one.

Then, how do you reel the raw silk from the cocoon and what kind of thread can you make? Let us introduce the outline from the perspective of the wearer.

古代人们在自然界中收集了各种茧，并通过智慧和创造力将它们制成线。自5,000年前以来，成功地驯化了蚕，将就可以稳定地收集大量的线。然后在明治时代，以富冈丝绸厂为代表的工厂开始生产生丝。

尚未驯化的各种野生蚕茧，根据类型的不同，不能制成生丝，有些茧只能絹丝（纺丝）。即使到现在，熟练的技术人员要逐个地处理纺丝。然后，您如何从茧子上取生丝，就可以制成什么样的线呢？ 知道它们的基本情况就让我们从穿着者的角度来介绍一下它们。

1. 生糸 / Raw silk / 生丝

生糸の表面をおおうセリシンは、乾燥するとガサガサして麻のようですが、精練の仕方ではシャリシャリにも柔らかくもなって、千変万化させる事ができます。野蚕の場合はセリシン層が薄いので、ほとんど精練しないで使っていることもあります。

生糸の太さは、デニールで表示されます。9000mの長さで1gの糸を、1デニールと呼び、標準的なカイコが吐く糸の太さは3デニールです。通常、生糸として流通する21D(デニール)の糸は21中と表示され、繭糸7本を引き揃えてあります。この表示は合成繊維など長繊維のすべてに使われています。

Sericin, which covers the surface of raw silk, looks rough and like hemp when dried, but depending on the degumming way, its texture can be changed to be either dry or soft, on conditions. In the case of In the case of wild silkworm, the sericin layer is thin, so we sometimes use its yarn without degumming.

The thickness of raw silk is displayed in denier. A 9000 m long 1 g thread is called 1 denier, and a standard silkworm spits 3 denier thread. Normally, 21D (denier) yarn that is distributed as raw silk is displayed as 21 medium, and 7 cocoon threads are lined up. This display is used for all long fibers such as synthetic fibers.

覆盖在生丝表面的丝胶在干燥时模上去沙沙的看起来像麻，但是随着脱胶的作用，它会变得松脆或柔软，并且可以千变万化。在野生蚕的情况下，丝胶蛋白层很薄，因此我们有时在不脱胶的情况下使用它。

生丝的厚粗细旦尼尔表示。长度为9000m的1g线称为1旦尼尔，标准蚕吐出3旦尼尔的线。作为生丝分布的普通 21D(旦尼尔)纱线显示为21中，并排列7条茧线。此显示适用于所有长纤维，例如合成纤维。

*生地を傷めないように毛玉を切るときには、鼻毛切りバサミが便利です。

*When cutting fiber balls so as not to hurt fiber, a nose hair scissors are useful.

*当切割纤维球以免伤害面料时，鼻毛剪刀非常有用。

2. 玉糸 / Doupion silk (also referred to as Tamaito) / 双宫丝 (也称为 Tamaito)

1匹のカイコがつくる繭は1つです。しかし時々、2匹が一緒になって1つの大きな丸形に近い繭を作ります。これを玉繭と呼び、繭から繰った2本の糸が絡んだ節のある生糸を玉糸と呼びます。玉糸は、シャンタンなど、節を特徴とする絹織物に利用されます。

以前、愛知県豊橋市はこの玉糸の産地でした。

Each silkworm makes one cocoon. But sometimes two worms together make a big single round shape cocoon. This is called Tamamayu (double cocoon), from which raw silk thread with two pieces entangled and having nodes is formed. That silk is called "Doupion silk". Doupion silk has been used for silk fabrics characterized by specific nodes such as "silk shantung". Formerly Toyohashi City, Aichi Prefecture was the production area of this Doupion silk.

每只蚕都形成一个茧。但有时，两个蠕虫会共同构成一个圆形的大茧。这就是所谓的双宫茧(双茧)，从该茧中缠结了两条生丝线，形成了多节的丝，称为“双宫绸”。双宫绸已经被用于“丝绸山东”的丝绸织物。

以前，爱知县丰桥市曾是这种双宫绸的生产地。

3. 真綿（まわた） / Floss silk (also referred to as Mawata) / 丝绵 (也称为 Mawata)

繭を、アルカリ溶液などでセリシンをとり除いてから柔らかくし、流水でよく洗い繭の上下の薄い方からサナギを取り除くと真綿になります。真綿をほぐしながら枠で平たく引き延ばすと、角真綿や袋真綿になります。

生糸を繰るのに不適切な玉繭や変形した繭、大きさの揃わない繭、切り繭などを真綿にし、真綿から糸を引けば、繭の全てが糸として使えます。

本来「わた」は、絹繊維の絡み合った塊のことでした。しかし、平安時代に日本に入り、江戸時代では各地で栽培されるようになった「木綿から作ったわた」の方が一般的になり、木綿の綿との区別のため、絹のわたを真の綿、「真綿」と呼ぶようになりました。

真綿は、昭和40年頃まで全国の布団商で販売され、布団を作るときに布との滑り止めと保温のため、木綿のわたを真綿で包みました。また、「どてらや半てん*」の保温とわた切れ防止用にも使われていました。現在では紬糸に利用されています。

After removing sericin from cocoon with an alkaline solution, etc. to soften cocoon, wash it thoroughly with running water and remove the pupa from the thin upper & lower parts of the cocoon to obtain Floss silk. While raveling, flatten silk using a frame to make square floss silk or floss silk bag. If cocoons that are not suitable for reeling raw silk, deformed cocoons, uneven size of cocoons or cut cocoons are made into Floss silk, and thread is pulled from it, whole cocoons can be used for the thread.

Originally “Wata” was designated in Japanese to entangled mass of silk fibers. However, since “Wata” made from cotton, which entered Japan during the Heian period and was cultivated in various places during the Edo period, became more common than silk Wata, and in order to distinguish silk Wata from cotton Wata, silk Wata is now called "Mawata (Floss silk)".

Floss silk had been sold at futon dealers nationwide until around 1965, and when making a futon, cotton Wata was wrapped with Floss silk for slip prevention against covering cloth and heat retention. It was also used to "DOTERA and HANTEN *" for heat retention and prevention of cotton disconnection. Nowadays it is used for hand spun silk.

用碱性溶液从茧中除去丝胶会使其软化，用流水将其洗净，然后从茧的上方和下方的薄壁上除去也蛹，将会丝绵。在放松茧的同时，如果使用框架将其弄平整以制作方丝绵或包袋丝绵。

将不适合做生丝的双宫茧，变形的茧，大小不等的茧或切茧制成丝绵，如果从丝绵中拉出线，则整个茧都可以用作丝线。

最初，日式丝“绵”是缠结的蚕丝纤维的块。但是，棉花制的“棉”在平安时代进入日本，并在江户时代在各个地方种植，因此变得更加普遍，并且为了与众不同来自棉花棉，真丝现在称为将是“真正的绵：真綿(Mawata)：丝绵”。直到1965年左右，丝绵才在全国的蒲团经销商处出售，制造蒲团时，棉花被包裹在丝绵中以防止其滑到并保温性更高。而且，至今它还被用了于保持“Dotera和Hanten*”的温度，并防止棉花散落。如今，它用于手纺丝纱。

4. 紬糸 / Hand spun silk thread (also referred to as Tsumugi thread) / 手纺丝纱：捻丝线(也称为Tsumugi线程)

真綿を引き伸ばして手で紡いだ糸や、精練したくず繭を足踏み機などの道具を使い、作った糸を「紬糸」と呼びます。

現在は、織物を楽しむ人々や創作作家が、織物に味わいをつけるために緯糸に使っています。

Threads made by stretching floss silk and twisting it by hand, or thread made from degummed waste cocoons with tools such as a foot stepping machine are called " Hand spun silk thread".

Currently, creative artists and people who enjoy hand-weaving in the workshop use it for weft to add flavor to the woven fabric.

通过拉伸丝绵并手工捻制而成的线，或通过脚踏机等工具使用脱胶的废茧制成的线称为“手纺丝纱：捻丝线”。当前喜欢织的人，与富有创造力的艺术家将其用于，以增加机织织物的风味。

---

*どてらや半てん：着物型で中わたが入ったものを指します。前が開いた形で、腰丈のものが半てんです。

*Dotera and Hanten are Kimono type coat with cotton between outer cloth and lining. Type with front open, and of short coat size is called Hanten.

*Dotera和Hanten：和服型，外层布料和衬里之间有棉质。前面开着的，短大衣尺寸为Hanten。

紬には高価な織物もありますが、和服の正式な絹糸は生糸とされ、紬糸にはくず繭なども使われるため、「紬織」は、高価であってもフォーマルな場所では着用しないのが通例です。

Though there are expensive textiles in Tsumugi, the formal silk thread for Kimono is considered to be raw silk. Since waste cocoons etc. are used for manufacturing of Tsumugi thread, Tsumugi ori is customary not worn at the formal places even if it is expensive.

Tsumugi也用于制作很昂贵的丝绸和服，和服的正式丝线是生丝，由于下脚料茧与废线也被用于作Tsumugi丝线习惯上即使它很贵也不穿在正式的場合。

5. 絹紡績糸(絹紡糸)、絹紡紬糸 / Spun silk yarn, Noil silk yarn / 绢丝(纺丝)、油丝

銘仙などの着物用や富士絹にも使う、繊維の長い生糸に近い高価格な糸から、安価な下着や靴下などに使う短繊維の糸まであり、様々な製品に使われ、綿やリネンなどとの混紡にも利用されています。絹紡糸の太さは木綿などと同様に番手で表示されます。番手表示は重さを一定にしてその長さで糸の太さを表示するもので、デニールとは逆に数値の多いものほど細くなります。

It has been used for various products, from high-priced threads that are close to long raw silk used for kimonos as Meisen, wide Fujiginu (spun silk-fabrics) to short-fiber threads used for inexpensive underwear and socks. It is also used for blending with cotton and linen. Thickness of the silk yarn is displayed in count (BANTE), as in cotton. Count display shows the thickness of the thread with the weight kept constant, and the larger the number is, the thinner becomes.

价格的接近生丝纱的长纤维的高价线也被用于各种产品，(如和服：铭仙绸和宽幅的富士丝：絹紡糸)到用于廉价的内衣和袜子的短纤维线。它也用于与棉和亚麻的混纺。线纱的粗细以支数显示，与棉一样。计数显示在重量保持恒定的情况下显示线纱的粗细，数字越大，线纱变得越薄。

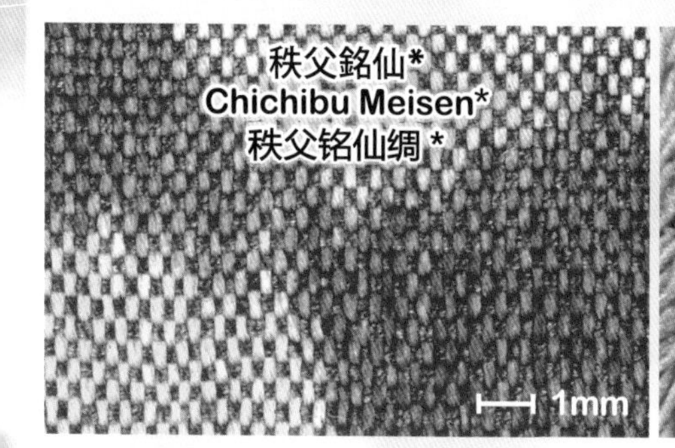

生糸の製造行程で残る、くず糸やくず繭を精練して真綿にし、不揃いな繊維を一定の長さに切って機械で紡いだ糸を絹紡糸と呼びます。

絹紡糸は、セリシンを除く程度により三分練り、五分練り、七分練り、本練り(完全精練)などに区分されます。

Waste silk yarn and waste cocoon, and non-standard cocoon, which are left over at the raw silk manufacturing process, are degummed and made into Floss silk. Irregular fibers are cut into a certain length, and spun with machine. That yarn is called Spun silk yarn.

Spun silk yarn is classified as follows:

30% degumming / semi (50%) - degumming / 70%degumming, and complete (100%) degumming (complete degumming) depending on the removal level of sericin.

绢丝纱是将在生丝制造过程中残留的废丝纱、非标准茧和下脚茧脱胶而成，并将不规则纤维剪切成一定长度后，通过机器纺成的纱。

根据丝胶的去除程度，将绢丝纱分为 30%脱胶，半脱胶，70%脱胶和所有脱胶(完全精练)。

低価格帯の下着
Low price range underwear
低价格范围内衣
1mm

絹紡糸の製造工程でできるくず(ブーレット：bourette)を紬糸紡績機で紡績した糸を絹紡紬糸と呼びます。

Noil silk yarn are threads that are spun with a pongee spinning machine using bourette.

油丝是使用绢丝纱纺丝过程中产生的短绒头(bourette)由紬丝纺绩机纺制的线。

＊秩父銘仙：秩父織塾横山工房株式会社。https://yokoyama-koubou.com

＊Chichibu Meisen: Chichibu Orijuku Yokoyama Kobo Co.,Ltd. https://yokoyama-koubou.com

＊秩父铭仙绸：秩父織塾横山工房株式会社。 https://yokoyama-koubou.com

Ⅳ. 絹の機能性利用 / Utilization of silk functionality / 丝绸的功能用途

約30年前、東京農工大学教授の故平林潔先生が「食べる絹」を発表されて以来、各研究機関で急速に絹の機能性が研究され始め、昨今では絹入り化粧品、石鹸、サプリメント、人工骨、下着等々活発な開発が進んできました。

家蚕の絹は、中心部の約75%が繊維としてのフィブロインで、表面から約25%は、にかわ質のセリシンです。産業的にセリシンは産業廃棄物として処理をする厄介もので、セリシンを除いたフィブロインのみが重要でした。

ところがその後の研究で、手間を掛けて捨てていたセリシンに機能性が高い事がわかり、医薬品など、医療関係への利用についての研究が盛んになりました。

About 30 years ago, the late Professor HIRABAYASHI Kiyoshi, a professor at Tokyo University of Agriculture and Technology, announced "Eating Silk", and many research institutes have started rapidly to study the functionality of silk. Nowadays cosmetics and soap with silk, nutritional supplement, artificial bone, underwear, and so on have been actively developed.

About 75% of raw silkworm silk consists of fibroin which exists at the center of fiber and its about 25% consists of sericin which is glue like material and exists at the surface of fiber. Only fibroin excluding sericin has been important industrially, and sericin has been unwanted material which has to be treated as industrial waste.

However, high functionality of sericin, abandoned so far, has been clarified in subsequent studies. And researches have been actively done about the use to medical applications including medicine.

大约30年前，东京农业技术大学教授已故的平林潔(HIRABAYASHI Kiyoshi)教授宣布了“食用蚕丝”的研究，每个研究机构都开始迅速研究蚕丝的功能，如今，化妆品和丝绸肥皂，丝绸补品，人造骨头和内衣的积极发展已经取得进展。

家蚕中心约有75%的是纤维蛋白纤维，表面约25%是胶粘型丝胶蛋白。只有纤维蛋白在工业上才是重要的，而丝胶作为工业废物是一个问题。

然而，在随后的研究中，原来被放弃，费时费力的丝胶具有很高的功能性，并且已经积极地研究了药物的使用和其他医学应用。

健康用品としての絹の利用 / Use of silk as material for health / 丝绸作为保健用品的用途

絹は世界中で高級衣料品として使われてきました。中世までは支配階級や富裕層、20世紀になってからは、お洒落な衣服として多くの人々が利用するようになりました。

しかし1960年代以降、ポリエステル(化学繊維)との混紡製品が、しわになりにくく速乾性があり、洗いやすいなどの機能性が高く評価され、半袖のワイシャツやコットンパンツが爆発的に売れました。現在、身の回りの繊維製品の主流は化学繊維との混紡製品になりました。

絹関連業者は、絹製品を水洗いし、スチームアイロンで仕上げますが、絹製品の洗濯表示はドライクリーニングです。

近年では全自動の洗濯機利用でも、アイロン不要の化学繊維の出現で、絹製品は、流通からどんどん落ちこぼれ、大手デパートでさえも洋装品に絹製品がなくなりました。

Silk fibers have been used for high quality clothes around the world. Until the Middle Ages, the ruling class and the affluent class used silk fibers for their high-end clothing and at 20th century, many people has begun to utilize them for their smart clothing. However, in the 1960s, blended products with polyester (chemical fiber) were praised for their functionality such as Wrinkle resistant, quick dry and easy-to-wash, and short-sleeved shirts and cotton pants were praised and distributed in large quantities. Nowadays, the mainstream textile products around us are mixed fiber products with chemical fibers.

Silk-related companies wash their silk products with water and use steam irons for surface treatment, but the laundry label for silk products is dry cleaning.

In recent years, the appearance of chemical fibers that do not require ironing even after being washed in a fully automatic washing machine has caused the silk products to drop more and more from distribution, and even at major department stores, silk clothing products have disappeared.

丝绸已被用作世界各地的高端服装。直到中世纪，统治阶级和有钱人都使用它，从20世纪开始，它被许多人用作流行服装。

但是，在1960年代，与聚酯(化学纤维)混纺的产品因其无皱纹，速干和易于洗涤等功能而受到赞誉，短袖衬衫和休闲裤也受到了好评和大量销售。如今，我们周围的主流纺织产品是化学纤维的混纺产品。与丝绸有关的公司用水洗涤丝绸产品，并用蒸气熨斗进行表面处理，但是丝绸产品的洗衣标签是干洗。近年来，即使在用全自动洗衣机洗涤后也不需要熨烫的化学纤维的出现，导致丝绸产品的经销越来越小，丝绸服装甚至在主要百货商店也消失了。

*洋装用絹織物を縫うための、絹のミシン糸も手芸品店などから消え、現在では工業用のコーン巻き(右ページ)も見かけなくなりました。

*The silk sewing thread used to sew silk fabrics for Western clothing has disappeared from handicraft stores, and nowadays, industrial corn rolls (right page) are no longer found.

*用来缝制西服丝绸的丝绸缝纫线已经从手工艺品商店中消失了，如今，不再有工业用圆锥卷(右页)了。

衣料品売り場に絹製品がなくなったばかりではなく、常設の服地売り場からも絹織物は消え*、20世紀の終わり頃まではフォーマル用として置かれていた、シルクツイードやシルクシャンタンなども昨今では見かけません。

21世紀になってから20年以上が過ぎ、絹製品の販売不振は一般的となり、若年層に至っては、着物以外に絹製品の存在を知らない時代となりました。しかし、美容、健康、機能性などの言葉への反応は、若年層が敏感です。

今後はシルク関連企業と連携し、若い世代と話す機会をつくり、絹製品がもたらす健康についてを伝え、日本の養蚕業を支える新しい蚕業を創出する活動へと繋げてゆきたいと思います。

Not only silk products in the clothing section, but also silk fabrics in the permanent clothing section* have disappeared, and silk tweed and silk shantung, which had been used for formal clothing until the end of the 20th century, are no longer seen recently.

More than 20 years have passed in 21st century, and sluggish sales of silk products have become commonplace, and now young people don' t know the existence of silk products other than kimono. However, they are very sensitive to the words such as beauty, health, and functionality.

In future, we would like to work with silk-related companies, have opportunities to talk with the younger generation to tell health brought about by silk products, and promote the activities to create a new sericulture industry that supports Japan's sericulture industry.

不仅在服装部分没有丝绸产品，而且织物柜台＊中也没有丝织物，到20世纪末一直用于正式服装的丝花呢和丝绸山东都不再可见。

在21世纪已经过去了20多年，丝绸产品的销售低迷已经成为现象，这是年轻人不了解和服以外的其他丝绸产品存在的时代。但是，年轻人对诸如美丽，健康和功能性等词语的反应更加敏感。

将来，我们希望与专门从事丝绸相关公司合作，以与年轻人对话丝绸产品带来的健康，并与各种活动联系起来，以创建支持日本蚕桑业的新型蚕桑业。

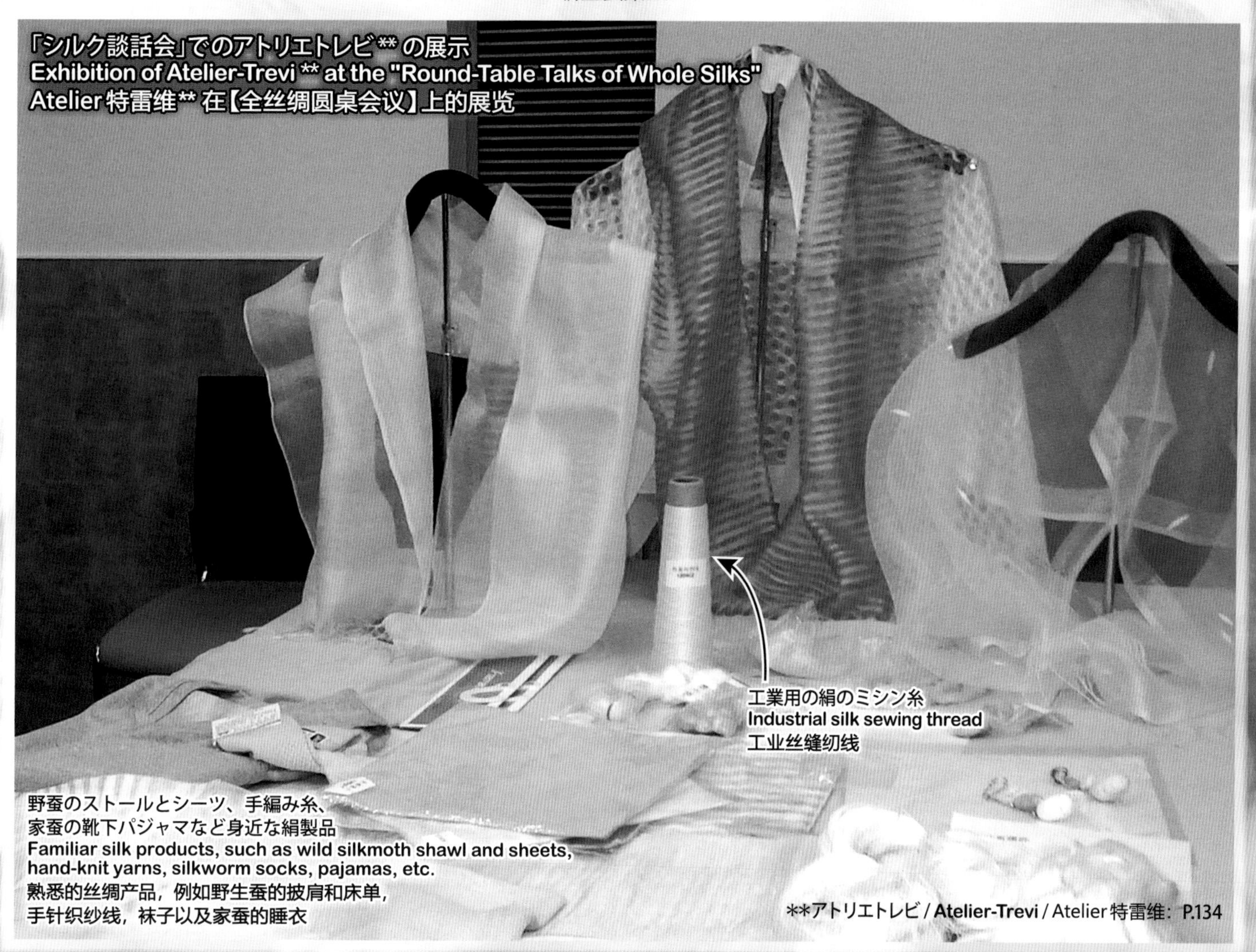

「シルク談話会」でのアトリエトレビ**の展示
Exhibition of Atelier-Trevi ** at the "Round-Table Talks of Whole Silks"
Atelier 特雷维** 在【全丝绸圆桌会议】上的展览

工業用の絹のミシン糸
Industrial silk sewing thread
工业丝缝纫线

野蚕のストールとシーツ、手編み糸、家蚕の靴下パジャマなど身近な絹製品
Familiar silk products, such as wild silkmoth shawl and sheets, hand-knit yarns, silkworm socks, pajamas, etc.
熟悉的丝绸产品，例如野生蚕的披肩和床单，手针织纱线，袜子以及家蚕的睡衣

**アトリエトレビ / Atelier-Trevi / Atelier 特雷维：P.134

ニューシルクロード「自然に学ぶものづくり」
New Silk Road "Manufacturing technology to be learnt from nature"
新丝绸之路【学习自然技术的社会实施工程】

長島 孝行 / NAGASHIMA Takayuki / 长岛 孝行

東京農業大学 農学部 デザイン農学科 科長・教授
Department Dean / Professor, Department of Design Agriculture, Faculty of Agriculture, Tokyo University of Agriculture
东京农业大学 农学部 設計农业系 主任 / 教授

シルクタンパク質の可能性 / Potential of silk protein / 丝蛋白的潜力

私は、昆虫などの小さな生き物の機能性を研究・応用し、そこから生まれたシステムを社会実装に近づけるインセクト・テクノロジーを提唱し、20年前から研究してきた。シルクを研究して驚いたことは、シルクタンパク質自体の持つ可能性である。

たとえば、繊維を構成するシルクタンパク質は、皮膚がんの原因となる紫外線のB波をカットし、菌の増殖を防ぎ、臭いや脂の吸着、保湿・保温性に優れ、難消化性を備えている。これが昆虫が繭を作る意味であり、究極のシェルターと考えることができる。また、パッチテストの結果、生体親和性にも優れていることもわかった。

I have been studying and applying the functionality of small creatures such as insects, advocating insect technology that brings the system created from it into closer to social implementation, and have been doing research on it for 20 years. Through my study on silk, I found the possibility of silk protein itself with surprise.

For example, the silk protein that make up the fiber has excellent properties, such as blocking ultraviolet B that causes skin cancer, prevention of bacteria growth, absorption of odors & fats, retention of moisture & heat and indigestible nature. This is the meaning that insects make cocoons, which can be thought of as the ultimate shelter. In addition, as a result of the patch test, silk protein also proved to be superior in biocompatibility.

我一直在研究和应用昆虫等小型生物的功能，并倡导昆虫技术，使由此创建的系统更接近于日劳生活中的应用，并且已经进行了20年的研究。当我研究丝绸时，令我惊讶的是丝绸蛋白本身的潜力。

例如，构成纤维的蚕丝蛋白能阻止引起皮肤癌的紫外线B波，阻止细菌生长，吸附异味和油脂，保留水分和热量，并具有也難的消化率。这就是昆虫制造茧的含义，可以将其视为最终的庇护所。另外，作为斑贴试验的结果，还发现其具有优异的人体親和性。

カイコの機能性を社会実装する / Social implementation of silk protein functionality / 丝绸蛋白功能在日劳生活中的应用

脂を吸着し難消化性であることは、メタボに効く「ダイエットサプリ」となり、赤ちゃんがお母さんの顔をなめても、安全な生物素材由来の「UVカットなどの機能性を持つコスメ原料」になる。

特筆すべきは、シルクは97%がタンパク質でできていて、特にフィブロインタンパク質は、水溶性、ゲル、パフ、フィルムなど様々な形状に加工できるため、極めて利便性が高い。

現在私の研究室由来の、食べるシルク、塗るシルクなど数種の製品が社会実装されている。

Owing to both functions of silk protein to absorb fat and to be indigestible, it can be a "diet supplement" that works against metabolic syndrome. It can be also a safe animal-origin cosmetic ingredient with the functionality of UV protection, with which safe cosmetics even in case that baby licks mother's face can be made.

Notably, 97% of silk is protein, and especially fibroin is particularly convenient because it is water-soluble and can be processed into various shapes such as gel, puff, and film.

Currently, several kinds of products such as edible silk and cosmetics silk from my laboratory are socially implemented.

吸附油脂及難的消化，这一事实使其成为对抗代谢综合症的【饮食补充品】。即使您的宝宝舔母亲的脸，它也是一种安全又可靠的生物成分，其源于成的为【具有抗紫外线功能的化妆品成分】。

值得注意的是，蚕丝是97%的蛋白质，因为它的水溶性的，因比，丝蛋白则特别方便加工成各种形状，例如凝胶，粉扑和薄膜。

目前，我实验室的食用蚕丝和化妆品蚕丝等几种产品已在市面销售。

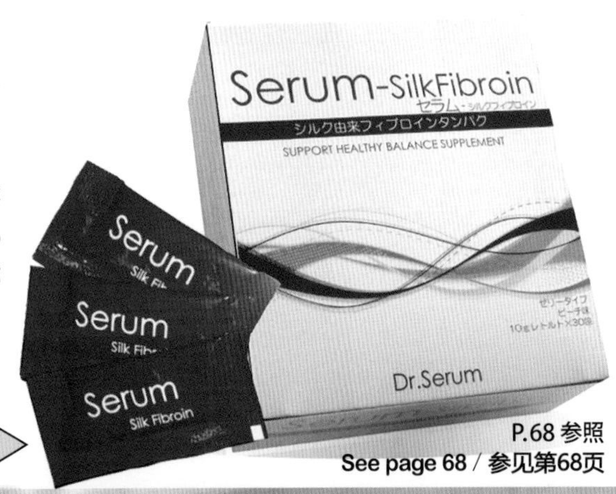

食べるシルク
Edible silk / 食用丝

P.68 参照
See page 68 / 参见第68页

画像提供：ドクターセラム㈱
Image provided: Dr. Serum co., Ltd.
图片提供: Dr. Serum co., Ltd.

第二次養蚕業の始まり / Second beginning of sericulture industry / 第二次养蚕业的开始

日本発「ジャパン-シルク」：昆虫工場
"Japan-silk" originating from Japan: Insect factory / 起源于日本【Japan-Silk】: 昆虫工厂

2015年、東京農業大学と連携した新潟県十日町市の㈱きものブレインは、松原 藤好 京都工芸繊維大学名誉教授の監修下で、高品質のシルク糸の大量生産を目指し、無菌・完全人工飼料によるカイコの飼育に成功した。

この飼育法では、純白な生糸を紫外線下に放置しても黄変しにくい。

無菌室内で、桑の葉パウダーを主原料とする人口飼料での全齢飼育は、掃き立てから営繭まで3回の給餌で病気も起きない。理論上では365日、毎日養蚕をすることも可能な、極めて生産性の高い養蚕であるが、人口飼料のコストが高く、従来の家蚕と比較すれば割高となる。

In 2015, Kimono Brain Co., Ltd. in Tokamachi City, Niigata prefecture in collaboration with Tokyo University of Agriculture, be under the supervision of Emeritus Professor of Kyoto Institute of Technology MATSUBARA Fujiyoshi, they aimed mass production of high-quality silk thread, succeeded in raising silkworm by sterile and complete artificial feed. With this raising method, pure white raw silk does not easily turn yellow, even if it is left under ultraviolet light.

During the length of the larval period in a sterile room, artificial feed made mainly from mulberry leaf powder caused no illness by three times feeding from beginning of silkworm rearing to cocooning. Theoretically, 365 days sericulture can be possible and the productivity looks very high, but that system is not cost competitive compared with the conventional sericulture because of high artificial feed cost.

2015年，新泻县十日町市的Kimono Brain株式会社与东京农业大学合作，在京都工艺纺织大学名誉教授MATSUBARA Fujiyoshi的指导下，旨在批量生产高质量的丝线，用完全无菌的人工饲料成功饲养了家蚕。用这种培育方法，即使将纯白色的生丝放在紫外线下，也不容易变黄。

在无菌室中，以桑叶粉为主要原料的人工饲料进行全龄育种，扫蚕（蚕种开始）到結茧仅三次喂养就不会引起疾病。从理论上讲，可以365天养蚕，产养蚕，但是人工饲料的成本很高，与传统的蚕相比，价格昂贵。

フラボノイドが豊富な「みどり繭」*
"MIDORI MAYU" rich in flavonoids*
富含类黄酮的 "MIDORI MAYU" *

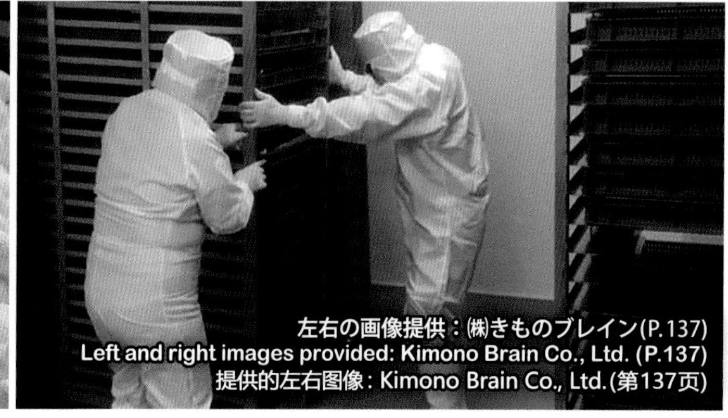

左右の画像提供：㈱きものブレイン(P.137)
Left and right images provided: Kimono Brain Co., Ltd. (P.137)
提供的左右图像：Kimono Brain Co., Ltd.(第137页)

すでにカイコからはペット用のワクチンなどが生産されているが、今回の技術開発は無菌養蚕であるため、用途範囲がずっと広がる可能性を持つ。初年度は2トン、翌年は5トン生産し、数年後には数十トンを目指す計画である。

Vaccines, etc. for pets have already been produced from silkworms, but since the technological development this time is aseptic sericulture, it has the potential to extend the range of applications. We plan is to produce 2 tons in the first year, 5 tons in the next year, and several dozen tons in a few years.

宠物等的疫苗已经由蚕生产，但是这次的技术发展是无菌蚕，有望扩大应用范围。我们计划第一年生产2吨，第二年生产5吨，几年后生产几十吨。

伝統的な養蚕：シルバーかいこ、ちびっ子かいこ、お家でかいこ、訪問かいこ、一億総活躍へ
Traditional sericulture: Elderly Sericulture, Kids Sericulture, Home Sericulture, Visiting Sericulture, and Engagement of All Citizens
传统养蚕：老年人蚕茧，儿童蚕茧，家庭蚕茧，来访蚕茧，全民参与

一方、屋内での古典的繭作りにも新しい方向が生まれた。

On the other hand, a new direction has also been created for indoor traditional sericulture.

另一方面，室内传统蚕桑创造了新的方向。

＊みどり繭の色：カロテノイドとフラボノイドが豊富な、黄色い繭の品種。黄色味が深い。

＊MIDORI MAYU color: It is a variety of yellow cocoon rich in carotenoids and flavonoids. Yellow is deep.

＊MIDORI MAYU的颜色：它是多种黄色的茧，富含类胡萝卜素黄酮，黄色很深。

これまでも、市民養蚕と呼ばれる活動が各地で展開されていたが、注目すべき活動が介護施設や、シルバーの方々を中心に実施されている。

注目されているのは、「訪問かいこ」。

介護施設が中心となり、その人の体調に合わせて養蚕をし、在宅高齢者や障がい者の自立を引き出すことを目的としたサービスである。

完成した大小様々な繭は、化粧品原料として非繊維利用され、わずかでも利益が配分される。

Until now, an activity called civic sericulture has been carried out in various places, and notable activities are carried out mainly in nursing homes and elderly people. What is attracting attention is "Visit Sericulture". Service is centered on nursing care facilities and aims to sericulture according to the person's physical condition and bring out the independence of the elderly at home and people with disabilities. Completed cocoons of various sizes are used not for fiber but for material for cosmetics, and the profit is distributed even if it is small.

迄今为止，已经在各地进行了名为“公民蚕桑业”的活动，但值得注意的活动主要是在养老院和老年人中进行的。吸引人们注意的是“来访蚕茧”。

服务以护理设施为中心，旨在根据人的身体状况进行蚕桑服务，并实现在家中老年人和残疾人的独立性。各种尺寸的成品茧被用作化妆品的非纤维材料，即使是最低生产也可获得利润。

各家庭に配る桑を、鉢植えで大切に育てる
Mulberry to be distributed to each home is grown in a pot carefully.
盆栽桑，精心养育并分发给每个家

施設の桑畑、桑は挿し木で増やす*
Mulberry field in the facility, mulberry is increased with cuttings*
设施中的桑树田，用插条增加桑*

このサービスの背景には、養蚕への新たな考え方がある。カイコの繭糸を絹織物として利用するには、太すぎたり割れたりした規格外の繭は廃棄するので歩留まりが悪い。シルクタンパクを利用するなら、きれいな形の繭でなくても一向に構わない。羽化した後の繭でも大丈夫。むしろ糸としては使えない崩れた繭のほうが材料としては処理が楽である。だから、かつての養蚕より飼育を簡単にし、糸を繰らないために、厳密な管理は不必要である。

施設側には、バリアフリーの反対のコンセプトで7割の利用者が改善しても、家に帰ると体を動かさなくなり、また悪化する悩みがあった。

そこで、自宅で数百匹のカイコを飼ってもらう。

Background to this service is a new way of thinking about sericulture. When silkworm cocoon thread is used for a silk fabric, cocoons out of specification (too thick, cracked, etc.) are discarded, that is, the yield is not high. In case silk protein is utilized, uniformly shaped cocoons are not needed at all.

Cocoons after emergence are also okay. Rather, the broken cocoon, which cannot be used as a thread, is easier to process as a material. Since it is not to be used as raw silk, its breeding method is simpler than the previous sericulture and does not require strict management. The facility side had the problem that even if 70% of users improved with the opposite concept of barrier-free, they could not move their body when they returned home, and it would worsen. Therefore, we asked them to keep several hundred silkworms at home.

该服务的背景是思考蚕桑文化的新方法。为了将蚕茧线用作丝织物，要将太粗或破裂的非标准茧丢弃，导致单产降低。如果您使用丝蛋白，则无需形状精美的茧。出蛾后的茧也可以。相反，不能用作纺线的，破裂的茧更容易加工为材料。由於不要用作于生丝，因此，其培育方法比以前的蚕桑更简单，并且不需要严格的管理。

设施方面的问题是，即使 70%的用户使用了相反的无障碍概念，进行了改善，回到家之后也无法活动自己的身体，而且还会荟来身体状况恶化的烦阂。因此，我们要求他们在家中饲养几百只蚕。

*画像提供（同）六次産業。
https://www.facebook.com/RokujiSangyou/

*Image provided Rokujisangyo LLC.
https://www.facebook.com/RokujiSangyou/

*提供的图片 Rokuji产业 LLC.
https://www.facebook.com/RokujiSangyou/

カイコの世話で知らぬ間に身体を動かし、リハビリができてセラピーにもなる。その間、施設の桑畑からカイコの餌になる桑の葉を届けることで、施設側に生活状況がわかり、健康的な日常への更なる支援ができる。

そしてわずかであっても、利用者の収入につながれば生活の質が向上し、新たな社会的な価値が創出できる。

While caring for the silkworm, he moves his body insidiously and can rehabilitate himself, which can be a therapy. Meanwhile, by delivering the mulberry leaves for Silkworm bait from the mulberry field at the facility, the facility side can know his living conditions at home and offer further support for healthy daily life. And, even if the amount is small, it will improve the quality of life and create new social value if it is connected to the user's income.

照顾蚕，您可以在不知不觉中活动自己的身体，并且可以康复并成为一种疗法。同时，通过从该设施的桑园收获和运送桑叶，也可以了解生活条件并进一步支持健康的日常生活。而且，即使金额很小，只要有收入，能改善生活质量就能创造新的社会价值。

第三のシルク：ワイルドシルク / Third silk: Wild silk / 第三丝：野生丝

これまでのように、桑で育てあげたシルクを第一のシルクとすれば、昆虫工場のシルクは第二のシルク、カイコ以外のシルクは第三のシルクである。シルクを吐く生物は、カイコのほか全てのガの仲間、チョウの仲間、コウチュウ類、クモ類、ダニ類など 10万種以上と私は推測している。中でもヤママユガ科の作る繭は、全ての紫外線(UV-A,B,C)波長をほぼ100%カットする。繭のサイズ、色にも多様性があり、金色、銀色、銅色のものまである。

インドネシアのクリキュラという蛾はアボガドの木を食い荒らす害虫として駆除されていた。しかしこの蛾はシルクで黄金の繭を作る。ならば金の糸ができそうなものだが、糸にすると金色になりにくい。

私たちが生産国に提案したのは、繭をそのまま広げてデンプンで貼り合わせ、天然黄金シートを作ること。プレゼンテーションとして愛知万博(2005年)の、千年共生村パビリオンの外壁の壁紙に使った。金色の外観が派手だったので覚えておられる方もあろう。

今では黄金繭を使ったランプシェードや金箔などがインドネシアで作られ、地場産業になっている。

If the silk grown in mulberry so far done is called the first silk, the insect factory silk is the second silk, and the silks other than silkworm are the third silk. I suppose that there are more than 100,000 species of organisms which eject silk, including all moth companions, butterfly companions, Coleoptera, spiders and mites other than silkworms. Above all, the cocoons made by the family Saturniidae cut almost 100% of all ultraviolet ray (UV-A, B, C). The size and color of the cocoons are also diversified, including some golden, silver and copper colors.

Indonesian moth called Cricula has been exterminated as a pest that devours avocado trees. But this moth makes the golden cocoon with its silk. So, it seems that golden thread can be made, but if thread is made, it does not easily turn golden.

What we proposed to the country of moth origin is to spread cocoons as they are and stick them together to make natural golden sheets. As a presentation, I used it as a wallpaper on the outer wall of the Millennium Symbiosis Village Pavilion at the Aichi Expo (2005). Someone may remember that the golden appearance was flashy.

Today, lamp shades and gold foils using golden cocoons are made in Indonesia and has become the products of local industry.

和以前一样，如果用桑树培育的蚕丝是第一的蚕丝，昆虫工厂培育的蚕丝是第二蚕丝，非蚕生産的蚕丝就是第三蚕丝。我推测，其中包括蚕，所有蛾类同伴，蝴蝶同伴，鞘翅目，蜘蛛和螨虫生物，有超过100,000种生物体能吐出蚕丝。最重要的是，由天蚕蛾科(Family Saturniidae)制造的茧几乎切断了所有紫外线(UV-A,B,C)波长的100%。茧的大小和各不相同颜色也多元化，包括有东西金色，银色和铜色。

印尼飞蛾 Cricula 是一种吞噬鳄梨树的害虫，而被除灭。但是，这种蛾子吐出了构成黄金茧的丝绸。如果是这样，似乎可以制作金纱线，但是如果将其制成纱线，它就不容易变成金色。

我们向原产国提出的建议是按自然散布茧，然后将它们粘在一起以制成天然金片。作为演示，我将其用作爱知世博会(2005年)千年千村馆外墙上的墙纸。有人可能还记得金色的外观很豪华。如今，灯罩和用金茧制成的金箔在印度尼西亚生产，正在成为当地的产业。

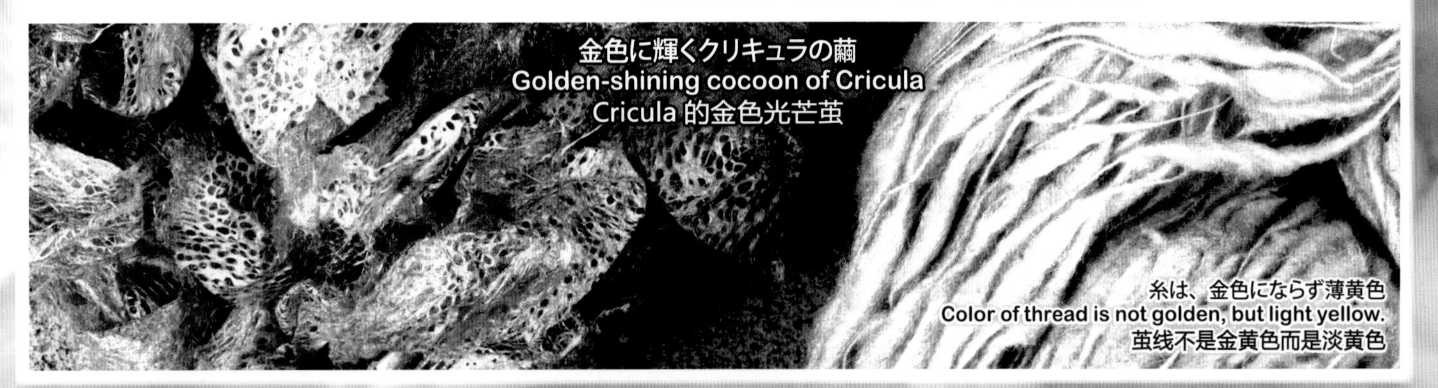
金色に輝くクリキュラの繭
Golden-shining cocoon of Cricula
Cricula 的金色光芒茧

糸は、金色にならず薄黄色
Color of thread is not golden, but light yellow.
茧线不是金黄色而是淡黄色

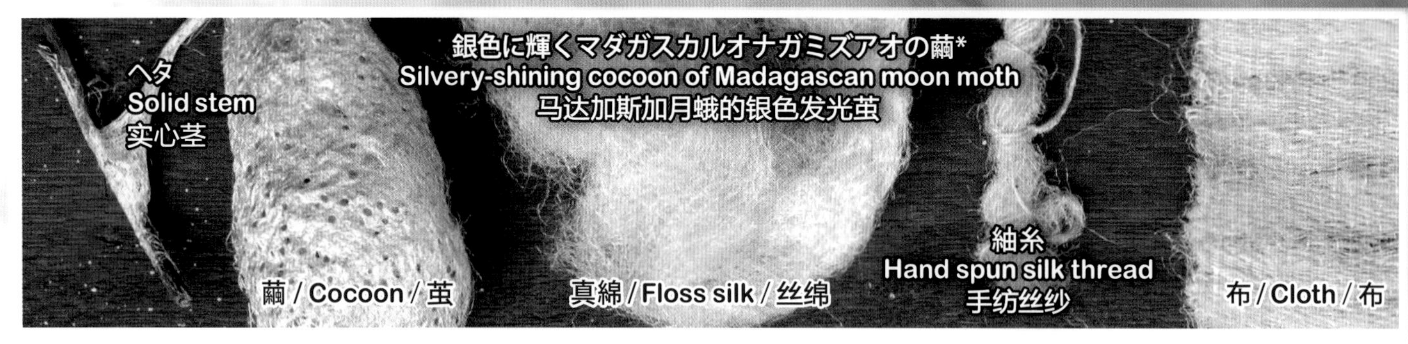

また、マダガスカルに生息するマダガスカルオナガミズアオという蛾は、シルクで銀色に輝く繭(まゆ)を作る。しかし成虫の羽化後、穴の開いた繭(出殻繭(でがらまゆ))を年間3kg程度採取するが、生きた繭や成虫を採集すると絶滅の可能性がある。

そこで、出殻繭は糸にせず、高価値のジュエリーに仕立て、ファッションショーでモデルに着用してもらった。

このように従来の蚕業の発想、カイコの出すシルクを、いかにすぐれた繊維に仕立て上げるか、いかに生産性を高めるかなどということから離れて、昆虫の出すシルクの機能そのものに着目してみると、今まで顧みられなかった可能性が大きく拓けることがわかる。

現在の日本に、新しい機能性繊維の開発は重要である。なぜなら衣食住から見れば、日本は「食」より「衣」の自給率が格段に低く、綿も絹も化学繊維もすべて外国産である。

ここまで述べてきた第三のシルク(野蚕シルク)は、熱帯地方に生息するものが多く、生産数にも限りがある。しかし環境に優しい機能性繊維としての開発は、家蚕の需要も喚起し、第三のシルクを昆虫工場で生産する未来も拓かれるであろう。

カイコガ科のカイコのように、ヤママユガ科で唯一家畜化されたエリサンは、標準的なカイコと同程度の太さの繭糸だ。日本でも1940年代には飼養されていたが、その後使われることは少なかった。それは、エリサンの繭は最初から穴が開いた破風抜け繭(はふぬけまゆ)で、生糸が繰れないからだ。

In addition, the Madagascan moon moth, which lives in Madagascar, makes silk cocoons that shine in silver. However, after adult emergence, about 3 kg of cocoons with holes are output per year, but if live cocoons and adults are collected, they may be extinct. Therefore, I didn't use the cocoon shells as threads, but made them into high-value jewelry, and had them be worn by models at fashion shows.

Separate from the traditional sericulture concept of silk production, that is, how to produce excellent silk from silkworm and how to heighten its productivity, many possibilities which have been overlooked can be found if we focus on the function of silk itself.

Nowadays, development of new functional fibers is important to Japan. This is because the self-sufficiency rate of Japan is far lower in "Clothing" than in food. Cotton, silk, and chemical fibers are all produced abroad.

Many of the third silks (Wild silk) described so far live in the tropical area, and their production volume is limited. However, development of silk as an environmentally friendly functional fiber will stimulate demand for silkworms, and will open the way to produce third silks in the insect factory in future.

Like the silkworm of the Bombyx mori family, Eri silkworm, the only domestic animal in the Saturniidae family, produces a cocoon filament of the same thickness as that of the standard silkworm. In Japan, it was raised in the 1940s, but few have been used afterwards. That is because Eri silkworm's cocoon is a cocoon on which a hole is opened from the beginning (Thin-end cocoon), so raw silk can' t be reeled.

此外，生活在马达加斯加的马达加斯加月蛾使丝茧发出银色光芒。然而，成年后，每年能收集约3公斤的穿孔茧(壳茧)，但如果收集活茧或成虫，则有灭绝的可能。因此，我们非将茧壳用作线，而是将它们制成了高价值的時装珠宝，并让模特们在时装秀上穿戴它们。

这样，与传统的蚕桑蚕丝生产理念不同，即如何从蚕中生产优质蚕丝以及如何提高蚕丝的生产率，如果我们专注于蚕丝本身的功能，就会发现许多被忽视的可能性。

在当今的日本，重要的是开发新的功能性纤维。这是因为从日本的“衣”的自给率来看比食品低得多，而且棉，丝和化纤原材料都产于国外。

到目前为止，描述的许多第三种丝绸（野生蚕）都生活在热带地区，其产量有限。然而，作为一种环保的功能性纤维的发展将刺激对蚕的需求，并为在昆虫工厂生产第三种丝绸开辟未来。

与家蚕家族(蚕蛾科)的家蚕一样，由天蚕蛾家族的唯一家养动物 Eri silkworm(樗蚕）也是一种茧丝，其粗細度与标准家蚕相同。尽管它是1940 年代在日本繁殖的，但此后很少使用。这是因为樗蚕的茧从一开始就在其中有一个孔(薄头茧)，这使得无法生产生丝。

＊P.55、クリキュラの繭と糸。マダガスカルオナガミズアオの繭と真綿、糸と布は、ワイルドシルクミュージアム（P.117）所蔵。

*P.55, Cricula cocoon and its thread. Madagascan moon moth cocoon and its floss silk; thread and cloth are owned by Wild Silk Museum (P.117).

＊第55页，Cucula的茧和线。马达加斯加的月蛾茧和丝绵，线和布，拥有在野生丝绸博物馆(第117页)。

エリサンの繊維はカイコのシルクよりも柔らかく、またヤママユガの仲間のシルクには、繊維の中にナノレベルのチューブ状の穴が無数に存在することから軽く、UV遮蔽性が高く、温度・湿度調節に優れ、臭いなども吸着してくれる。

しかしこれだけだと細くて切れやすい。そこで綿やウールと同じように繭糸を短くカットし、長繊維に加工できる綿と混紡した新素材をシキボウ㈱と共同開発した。綿100%に比べ11%軽く、紫外線カット率は12%アップ。アンモニア消臭性も21%上昇しており、汗くさくならない。化学成分を加えない天然素材だから、肌の弱い乳幼児やアレルギー体質の人に最適で、しかも洗えるシルクなのでタオルやTシャツにも使える。

通常繊維に上記の機能を持たせるためには、繊維に薬剤塗布等が必要だが、この繊維は元々こうした機能特性を持っているため、その処理も必要なく、ベビー用途の使用が可能である。

この天然機能繊維を「エリナチュレ」と命名し、2013年にはキッズデザイン賞(P.59)、2016年にはグッドデザイン賞を受賞した。

この繊維からは以下の製品が販売されている。

Eri silkworm's fibers are softer than silkworm's silk, and Saturniidae's family silk has a lot of nano-level size tube-shaped holes in the fibers, so it is light and has function of high UV shielding, excellent control of temperature and humidity and odor absorption. But this is so thin and easy to be cut. Therefore, we jointly developed with Shikibo Ltd. a new material blended with cotton that can be processed into long fibers by cutting the cocoon thread short like cotton and wool. It is 11% lighter than 100% cotton, and the UV cutting rate gets higher by 12%. Deodorant property against ammonia is increased by 21%, and sweaty smell can be prevented. Since it is a natural material without the addition of chemical components, it is suited to people with allergies as well as infants with sensitive skin, and since it is washable silk, it can also be used for towels and T-shirts.

Normally, in order to impart the above function to fibers, it is necessary to apply medicine to the fibers, but since these fibers originally have such functional properties, the chemical treatment is not necessary and they can be used to baby.

This natural functional fiber is named as"Eri-nature", which won the Kids Design Award in 2013 (P.59) and the Good Design Award in 2016. Currently, from this fiber, the following baby clothes are sold.

樗蚕的纤维比蚕的丝绸柔软，天蚕蛾科的蚕丝在纤维中有许多纳米级的管状孔，因此它很轻，它具有很高的紫外线屏蔽性能，出色的温度和湿度控制，并吸收异味极好。然而它较细长易切割。因此，我们以与棉花和羊毛相同的方式，将茧线切短，并与Shikibo株式会社共同开发了一种新材料，可和棉混纺，可加工成长纤维。占11%比100%棉轻，紫外线截断率提高12%。氨气除臭性能也提高21%，可去汗味。由于它是一种天然材料，不添加化学成分，因此非情适合过 敏人群以及皮肤较弱的婴儿，并且由于它是可洗的丝绸，因此可用于做毛巾和T恤衫。通常，为了赋予纤维上述功能，必须对纤维施加药物，但是由于这些纤维本来就具有这种功能特性，因此无需化学处理并且可以用于婴儿。

这种天然功能纤维被命名为“Eri-nachure”，并于2013年获得了儿童设计奖(第59页)，并于2016年获得了优良设计奖。目前，使用该纤维，销售以下婴儿服装。

画像提供：ベッタ (P.138)
Image provided: Bétta (P.138)
图片提供: Bétta(第138页)

シルクの機能性から学ぶものづくり
Manufacturing method to be learnt from silk function / 从丝绸的功能中学到的制造方法

シルクは繊維利用だけではなく、高純度のタンパク質は私たち人類にとって非常に都合がよく、驚くような機能性が近年発見されてきた。同時にゲルや液体になど、シルクタンパク質の加工技術も急速に進んできた。

Recently silk has been found very useful to us humankind not only in the usage as fiber, but also in its surprising functionality, owing to high purity protein. At the same time, processing technology to make gel and liquid from silk protein has been rapidly advanced.

丝绸不仅使用纤维，而且高纯度的蛋白质更适合于我们人类，且近年来发现了令人惊讶的功能。同时，诸如凝胶和液体之类的丝蛋白的加工技术已迅速发展。

2017年、シルクタンパク質を使用したマウス実験で、アトピー性皮膚炎の改善が見られたことを発表した。

シルクタンパク質の機能性を良質に利用すれば、防腐剤のいらない機能性美容液、メタボ対策用サプリメント、アレルギーの方でも安心して使用できるUVカットクリームなどなど、ものづくりのアイデアは盛り沢山。

In 2017, I announced that atopic dermatitis was improved in a mouse experiment using silk protein.

If you use the excellent functionality of silk protein, there are many ideas for manufacturing high level products such as functional beauty essence that does not require preservatives, supplements for anti-metabolic substances, UV cut creams that can be used safely even by people with allergies.

2017年，我宣布在使用丝蛋白的小鼠实验中观察到特应性皮炎的改善。

如果您以高质量也使用丝蛋白的功能，则有许多制造思路，例如不需要防腐剂的功能性美容精华，抗代谢物质的补充剂，抗紫外线隔离霜，即使有过敏的人也可以安全使用。

*ベビー用品、シルバー世代、アレルギーの方にも安心して着られます。但し、シルクアレルギーの方もあるので、気をつけてください。

* Clothes can be worn with confidence in case of baby, elderly generation and people with allergy. However, since some people are allergic to silk, care should be taken by them.

*婴儿用品，白银一代和有过敏症的人都可以放心地佩戴它。但是，有些人对丝绸过敏，因此请小心。

エリサンのエコサイクル「エリナチュレ プロジェクト」
Eri silkworm's Eco cycle : "Erinature Project" / 樗蚕的生态循环【Erinature项目】

食料増産・繊維原料の安定的な確保・農業支援・地球温暖化問題の解決
Food production increase, stable supply of raw textile materials, agriculture support and solution of global warming issues
增加粮食产量，确保稳定的纺织原料供应，支持农业，解决全球变暖问题

キャッサバは作付けが簡単で収量が多く、やせた土地でも栽培可能な熱帯低木。これまで廃棄処分されてきた葉で、エリサンの幼虫を育て、キャッサバの芋は、タピオカやポテトチップスなどの食品、飼料用でん粉の原料となる。

このプロジェクトは、カンボジアの農家の自主的発展を支援目的とし、プロジェクトにとって、繊維原料の安定的な確保が期待される。

Cassava is a tropical shrub that is simple to plant, has a lot of yield, and can grow even on a lean ground. With leaves that have been disposed of until now, Eri silkworm larvae can be bred, and from cassava tuber, food such as tapioca and potato chips and raw materials for feed starch can be made.
This project aims to support the spontaneous development of farmers in Cambodia, and through the project, stable supply of raw fiber material can be expected.

木薯是一种热带灌木，易于种植，高产，即使在贫穷的土地上也可以耕种。迄今已丢弃的叶子长出了樗蚕幼虫的繁殖，木薯被用来做木薯和薯片等食物的原料，淀粉也可做饲料淀粉的原料。
该项目旨在支持柬埔寨农民的自主发展，预计该项目也将确保稳定的纤维原料供应。

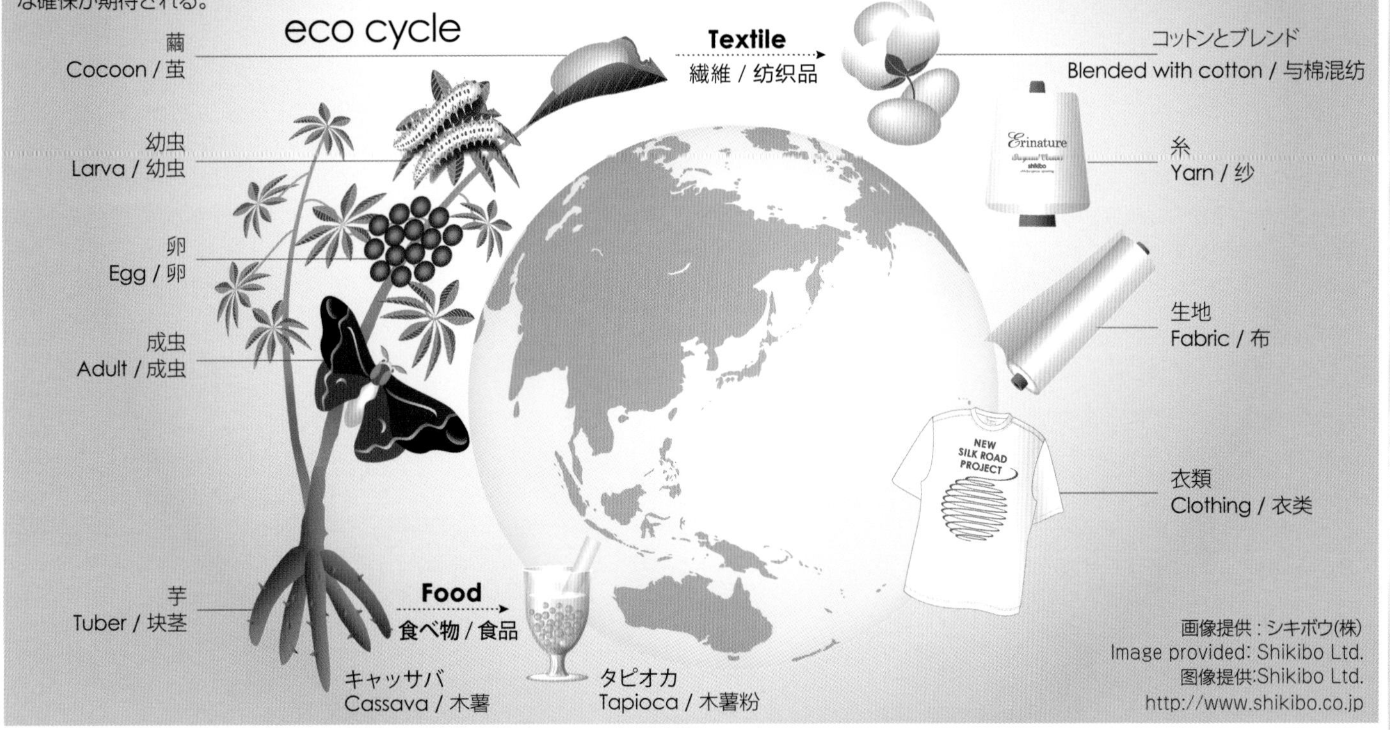

画像提供：シキボウ(株)
Image provided: Shikibo Ltd.
图像提供:Shikibo Ltd.
http://www.shikibo.co.jp

東京農業大学と、シキボウ㈱は、東南アジアの生産農家の支援・持続可能な産業創造などの観点から、カンボジアでエコサイクルを目指す、「エリナチュレ プロジェクト」を立ち上げ、衣料などに使用される生地(素材)を生産している。

2013年にキッズデザイン賞＊を受賞したエリナチュレは、プロジェクトのストーリーを賞賛された。この賞は子どもたちの安全・安心に貢献するデザイン、創造性と未来を拓くデザイン、そして子どもを産み育てやすいデザインの顕彰制度である。

Tokyo Agricultural University and Shikibo Ltd. have launched the "Erinature Project" aiming at an Eco cycle in Cambodia from the viewpoint to support farmers in Southeast Asia and create sustainable industries, and cocoon are producing fabrics (materials) used for clothing and so on.
Erinature, which won the Kids Design Award* in 2013, was praised for the story of the project. This award is a system that recognizes designs that contribute to the safety and security of children, designs that open creativity and the future, and designs that make it easy to give birth to and raise children.

东京农业大学和Shikibo Ltd.发起了"Erinature项目"，旨在从东南亚生产农户的支持，和可持续发展产业等方面的支持，以及从帮助樗蚕养蚕业的角度出发，在柬埔寨开展生态循环，茧正在用于服装生产等织物的材料。
Erinature 曾在 2013 年获得儿童设计奖*，所以和该项目有关的故事获得了赞誉。该奖项旨在表彰那些有助于儿童安全与安保的设计，为创造力和未来发展的设计，易于养育孩子的设计。

＊受賞作品には、「キッズデザインマーク」を使用できる。　＊"Kids Design Mark" can be used for the winning work.　＊【儿童设计标记】，可用于获奖作品。

伝統ある絹の文化を刷新・「ニューシルクロード」プロジェクト
Innovation of traditional silk culture: "New Silk Road" project / 改造传统丝绸文化:【新丝绸之路】项目

カイコの餌になる桑の葉にも新たな可能性が見つかった。桑という植物は、成長が早く無農薬で育つ。日本には300種以上の桑の品種があり、その中から「おいしく」、しかも有用な栄養素が多く含まれた品種を作り、「食用桑(NSRP 桑)」*と名づけた。ただの桑とは違い、色、香り、味などが優れ、管理も徹底している。また、血糖値を抑える物質も安定して含まれ、健康食品の青汁によく使われる大麦若葉と比較したところ、ビタミンA、カルシウム、鉄分などの含有量が比較にならないほど多いことがわかった。

5月に挿し木をすると、10月にはもう180㎝になるくらい成長が早く、この食用桑をブランド化すれば、高付加価値の桑の葉として農家の6次産業化に貢献できる。

2年かけて品種と製法にこだわった食用桑パウダーを、老舗の和菓子メーカーの「たねや」が「今までの桑の葉パウダーよりおいしい」と採用。ほかにもさまざまな用途で使われている。

New possibility was found in the mulberry leaf which is the silkworm bait. Plants called mulberry grow quickly without pesticides. More than 300 kinds of mulberry varieties in Japan, from which we created a "delicious" variety containing a lot of useful nutrients and named it "Edible mulberry (NSRP Mulberry)" *. Unlike general mulberry leaves, color, scent, taste, etc., are excellent and the control during cultivation is also thorough. In addition, substances that suppress blood glucose levels were stably included, and it was found that contents of vitamin A, calcium, iron, etc. were much higher than young barley leaves frequently used for healthy food green juice. When a cutting is made in May, mulberry trees grow very fast to about 180 cm in October, so by branding of this edible mulberry, it can contribute to the sixth sector industrialization of farmers as high added value mulberry leaves.

Taneya, a long-established Japanese confectionery maker, has adopted edible mulberry powder whose breed and manufacturing method have been intensively improved over two years, with the reason that it is more delicious than the mulberry leaf powder so far used.

在蚕的诱饵桑叶中发现了新的可能性。被称为桑树的植物生长迅速，无需使用农药。日本有300多个桑树品种，从中我们创造了一种“美味”的品种，其中含有大量有用的营养素，并将其命名为“食用桑树(NSRP桑树)”*。与普通桑叶不同，它具有出色的色，香，味，并且管理得当。而且，还具有稳定血糖水平的物质，此外，与经常用于保健食品绿汁的大麦幼叶相比，发现维生素A，钙，铁等的含量都很多。在五月份插枝的时，它会在十月份长到180厘米左右，如果将这种食用桑名牌化，它可以作为高附加值的桑叶，促进农民的第六次产业化。

两年来一直关注其品种和制造方法的食用桑粉被日本著名的糖果制造商“Taneya”采用，“比我们曾经使用过的桑粉更美味”。用于各种其他目的。

左右の画像提供：㈱たねや
Left and right images provided: Taneya Co., Ltd.
提供的左右图像: Taneya Co., Ltd.
http://taneya.jp

食用桑パウダーの主張は、食べた後にあなたの血糖値や鉄分、骨分に出る。

血糖値の気になるお父さん、カルシウムや鉄分の気になるお母さん、どなたにも利用可能で意外な効果、こんな素晴らしい健康食品の原料が、日本には昔からあった。

今後、桑は改めて注目されることは間違いない。

The effect of edible powder after taken appears on the level of glucose, iron and calcium in blood. To the father who has strongly concerned about blood glucose value and mother who has strongly concerned about calcium and iron content in blood, raw materials with such wonderful health effects are available in Japan for a long time. There is no doubt that Mulberry will regained attention in the future.

是食用桑粉后身体的血糖值较稳定，并迖到补铁，钙元素和骨骼质量中現出。关注血糖值的父亲，关注钙和铁含量的母亲，任何人都可以使用并显示有出乎意料的效果，如此出色的保健食品的原料在日本已经存在了很长时间。毫无疑问，桑树会重新受到关注。

＊食用桑(New Silk Road Project桑)農場の紹介：熊本常盤松シルクファームは、東京農業大学のOBによって2012年創立。約40aの農場に2000本の桑が生育しています。

＊Introduction of Edible Mulberry Farm (New Silk Road Project Mulberry): Kumamoto Tokiwa Matsu Silk farm was founded in 2012 by graduates of Tokyo Agricultural University. 2000 of mulberries are growing in the farm of about 40 ares.

＊食用桑树介绍(新丝绸之路项目桑树)农场：熊本Tokiwa Matsu丝农场是由东京农业大学OB于2012年建立的，在约40公顷的农场上种有2000桑树。

品種が同じ作物が、どこでも同様に育つわけではない。土壌条件、温度、照度などの環境によって成分がかなり異なるから、その土地に合わせて管理しなければならない。

我々は毎年桑の挿し木をするが、私たちの推薦するシルクファームは、熊本常盤松、南アルプス、福島県矢吹町*など、このほか日本各地で栽培が始まっている。

このように、カイコの餌となる桑の葉も使え、シルクは機能性食品や化粧品、新繊維になる。

私は、二千年の歴史がある養蚕業を、全く新しい形で再生し、シルクロード東端の日本から世界へ発信すべく取り組み、学生達のカリキュラムにとり入れ、若い世代が学び、引き継ぎ、改革し、活動を持続させて行くように、「ニューシルクロードプロジェクト」と呼んでいる。

Not the same crop plants will grow anywhere in the same way. Because components are quite different depending on the environment such as soil condition, temperature, illuminance, etc., control must be done to meet the each land's condition. We make a cutting of mulberry every year, and our recommending silk farms are in Kumamoto Tokiwamatsu, Minami Alps, Yabuki Town* in Fukushima Prefecture, However mulberry cultivation has already started in many areas around Japan. In this way, mulberry leaves can be used for purposes other than silkworm food, and silk becomes functional foods, cosmetics, and new fibers.

I have tried to revive the sericulture industry, which has a history of 2,000 years, in a completely new way and to transmit the result from Japan located in the eastern end of the Silk Road to the world. I call that project "New Silk Road Project" and intend for young generation to learn, take over, reform and continue its activity. Therefore I incorporate it into the curriculum of the students.

同一品种的作物,不会在任何地方都同样地生长。由于土壤条件,温度,照度等的成分因环境不同而有很大差异,因此必须根应地制宜地进行管理。我们每年都份插枝桑树,我们推荐的丝绸农场是熊本县常松,南阿尔卑斯山,福岛县的矢吹町*以及已经开始在日本其他地区种植。

这样一来,桑叶就可以用作蚕蛾诱饵以外的用途,而蚕丝可以用作功能食品,化妆品和新纤维。我将以全新的方式重现具有2000年历史的蚕桑业,并将其纳入学生的课程中,并我们试图其从丝绸之路东端的日本传到全世界,年轻一代自己学习和在接管,进行了改革并继续开展活动,我们称之为"新丝绸之路计划"。

*農事組合法人南アルプスシルクファーム
https://r.goope.jp/m-a-shiruku
矢吹町ニューシルクロードプロジェクト
http://yabuki-silk.jp/

*Agricultural Association Southern Alps Silk Farm
https://r.goope.jp/m-a-shiruku
Yabukicho New Silk Road Project
http://yabuki-silk.jp/

*南阿尔卑斯丝绸农场, 农业合作社
https://r.goope.jp/m-a-shiruku
矢吹町新丝绸之路项目
http://yabuki-silk.jp/

昆虫工場*から生まれる未来の絹
Future silk born from insect factory* / 生产来自昆虫工厂*的未来丝

これは、本大学の2017年度の学生募集用ポスター。毎年、大学が関わる研究の社会実装がポスターになる。

少子化傾向の日本において、食料生産を中心とした生産農学は、農学部にとって普遍的な研究テーマである。本学部では2017年度から、学際的な生物資源開発学科と、私が担当する先進的な作り方と、使い方のイノベーションを起こすデザイン農学科が加わった。「農」の知識は、今や機能性や加工を重視した医療や商品開発にも応用され、生き物の力をうまく利用すれば、人々の生活をより豊かにすることができる。

学生達が、「新しい農学」で未来の社会に貢献できるスキルを身につけられるように、私自身も様々に社会貢献に繋がる研究を続けている。

Photo above is a poster for 2017 student recruitment at our university. Every year, the social implementation of the research involving the university becomes a poster. In Japan, where the birthrate is falling, production agriculture centered on food production is a universal research theme for the Faculty of Agriculture. From 2017, the Department of Design Agriculture and the Department of Interdisciplinary Bioresource Development, which I am in charge of, have been added to this faculty. The former is, which intends to create innovation in advanced way of making and usage. Knowledge of "agriculture" is now applied to medical and product development that emphasizes functionality and processing, and if we use the power of living things well, we can enrich people's lives.

In order for students to acquire the skills to contribute to the future society with "new agriculture", I myself continue research to contribute to the society in various fields.

上图是该大学2017财年学生招募海报。每年，涉及大学研究的社会实施都成为海报。在人口出生率下降的日本，以粮食生产为中心的生产农业是农业学院的普遍研究主题。在本学部从2017年起，增加了跨学科生物资源开发部门，与增加了我负责的设计农业部门。该部门将是带来先进制造和使用方面的创新。现在，“农业”知识已应用于强调功能和加工的医学和产品开发中，如果我们充分利用生物的力量，我们就能丰富人们的生活。

为了使学生掌握用“新农业科学”为未来社会做出贡献的技能，我本人继续研究各种有社会贡献的研究。

---

*昆虫工場：㈱きものブレイン P.137

*Insect factory: Kimono Brain Co., Ltd. P.137

*昆虫工厂: Kimono Brain Co., Ltd. 第137页

シルクとゴミ、過去と未来
Silk and garbage, past and future / 丝绸和垃圾, 过去和未来

生糸を繭から引き出すときに、絡んだり切れたりする最初の糸をキビソと呼び、生糸を引いた後に薄い紙のように残る部分をビスと呼ぶ。生糸を大量生産した時代には、捨てられていたこともあったが、今では素朴な糸として使われたり、非繊維利用されたりしている。近年では、セリシンの特性が注目され、20世紀末までは捨てられていた農家毛羽も使われるようになった。

養蚕農家が激減し、作られる生糸も減り、くず繭もキビソも使う時代となり、さらに、非繊維利用が発達した結果、製糸関連の絹ゴミはほぼなくなった。

1935年のナイロンの発明と工業生産化以来、徐々に繊維が石油由来の化学合成の時代に変わってしまった。天然資源を駆逐した、石油製品は分解せず、海を漂い微細化し、エビや魚などを介して我々の口にも入る。私はこの持続性のないサイクルを止めたいと思い、20世紀の終わりころ、不要になった絹繊維をプラスチックのように成型してみた。芝に放置されても分解しないゴルフ ティーを、不要な絹で作れば、土壌分解してタンパク質を土に返す。マイクロプラスチックをあまり飲みたくないので、繭から透明なボトルも作ってみたが、当時はあまり注目されなかった。コストがかかるのだ。しかし、近年では化学合成繊維からのマイクロプラスチックが、海の汚染の主原因だと考えられてきた。

改めて人類は、生物圏に負担をかけない天然繊維、自然素材に最注目すべきである。

When reeling raw silk from the cocoon, the first thread that is entangled or cut is called Kibiso, and the part that remains like paper after pulling the raw silk is called Bisu. Although they were sometimes thrown away at the times of mass production of raw silk, they are now used as a rustic yarn, or it is used non fiber. In recent years, the characteristics of sericin have drawn attention and Farmer's cocoon floss that had been abandoned until the end of the 20 th century has also been used. Since sericulture farmers drastically declined and production volume of raw silk decreased, also use waste cocoon and Kibiso are now utilized. And as a result of the development of non fiber utilization, silk garbage including that in spinning process almost disappeared.

Since the invention of nylon in 1935 and industrial production start, gradually the fiber production has changed into the era of petroleum-derived chemical synthesis. Petroleum products that have destroyed natural resources will not be decomposed, and drift in the sea, miniaturizing, very fine particles enter body via shrimp and fish etc. I want to stop this unsustainable cycle, and at the end of the 20th century, I molded the waste silk fibers like plastic.

Plastic golf tees do not decompose when left on lawn, but if they were made of unnecessary silk, they would be decomposed and proteins would return to soil. I don't want to ingest micro plastic so much, so I tried to make a transparent bottle out of cocoons, but it didn't get much attention at the time. That's costs money. However, in recent years microplastics from synthetic fibers have been considered to be the major cause of sea pollution.

Again, humanity should pay the most attention to natural fibers and materials that do not impose a burden on biosphere.

从茧上 缫丝生丝时，缠结或切断得很好的第一条线称为条吐(Kibiso)，将生丝后残留得像薄纸一样残留的部分称为滞斗(Bisu)。在过去尽管有时将其丢弃在生丝的批量生产，时期，但现在已用作乡村纱，或用作非纤维。近年来，丝胶蛋白的特性引起了人们的关注，并使用起了直到20世纪末一直被废弃的斗农民的茧衣(Cocoon floss)。

养蚕业者急剧下降，制成的生丝减少了，因此下脚茧和Kibiso使用了，由于非纤维利用的发展，与纺丝有关的丝垃圾几乎消失了。

自从1935年发明尼龙并开始工业生产以来，纤维已逐渐转变为石油衍生的化学合成时代。破坏了自然资源的石油产品不会分解了，在海中漂流，微型化，将通过虾和将鱼等进入我们的嘴里。我想停止这个不可持续的周期，在20世纪末，我像塑料一样模制了废丝纤维。塑料高尔夫球座在草坪上放置时不会分解，但是如果用不必要的丝绸制成，则会分解，蛋白质也会重新回到土壤中。我不想喝太多的塑料，所以我试图用茧丝制成一个透明的瓶子，但是当时并没有引起太多关注。那要花钱。但是，近年来，来自合成纤维的微塑料被认为是造成海洋污染的主要原因。

重新，人类应该最关注不会给生物圈造成负担的天然纤维和材料。

プラスチック汚染の防止
Suppression of plastic pollution
抑制塑料污染

ゴルフショップで無料配布された竹のティー
Bamboo tees distributed free of charge at the golf shop
在高尔夫商店免费分发的竹子球座

固形になった絹繊維（学生の作品）
Solidified silk fiber (student work)
凝固丝纤维（学生作品）

シルクゲル*でおいしさを追求

Pursuing deliciousness with silk gel* / 用丝凝胶*追求美味

庄内の美味がたっぷり！「シルクのフルコース」

Plenty of delicious Shonai! "Full course of Edible silk in" / 有很多美味的庄内美味！【全程含丝蛋白课程】

グランド エル・サン（山形県鶴岡市）P.134
Grand el Sun (Tsuruoka, Yamagata) P.134 / Grand el Sun (山形县鹤冈市) 第134页

鶴岡のシルクと、庄内の旬の食材を使った、スペシャルメニューのフルコースは、イタリアンで人気のシェフの地元愛のたまもの。結婚式で、花嫁に美しく白い「ふわふわのシルクパン」を提供したいと考えたことが始まり。桑の葉のピュレとともにシルクの美味を提供しています。

A full-course special menu using Tsuruoka silk and Shonai's seasonal ingredients is a gift from a popular chef of Italian cuisine based on his love to the local area. At first, he wanted to offer beautiful white "fluffy silk bread" to the bride at the wedding ceremony. Offers the deliciousness of mulberry leaf puree and silk.

使用鹤冈丝绸和庄内的时令食材烹制的特别菜单全套餐，是一位受欢迎的意大利食物厨师的爱的礼物。在婚礼上，他开始想要为新娘提供美丽的白色"蓬松丝面包"吋它就开始了。提供桑叶原浆和蚕丝的美味。

JR羽越本線鶴岡駅より車で5分
5 minutes by car from JR Uetsu Main Line Tsuruoka Station
从JR上越本线鹤冈站乘车5分钟

チャペルとシェフのオリジナル料理が人気
The chapel and the chef's original cuisine are popular
小教堂和厨师的原创美食的人气

シルクのフルコースは特別メニューのため、1週間前までに予約が必要
"Full course with silk protein" is the special menu, so reservations must be made at least one week in advance
将特别菜单【全程含丝蛋白课程】至少提前一周预订

*シルクゲルは、同じ地域の松岡㈱製です。2021年現在日本の欧式器械製糸場は、群馬県安中市の碓氷製糸㈱とこの松岡㈱の2カ所です。

*Silk gel is manufactured by Matsuoka Co., Ltd. in the same area. As of 2021, there are two European-style Instrument silk mills in Japan, Usui Raw Silk Manufacturing Co., Ltd. and Matsuoka Co., Ltd. in Annaka City, Gunma Prefecture.

*丝绸凝胶由松冈株式会社在同一地区生产。 截至2021年，日本有两家欧式机器丝厂，分别是位于群马县安中市的臼井生丝株式会社和松冈株式会社。

シルク入りバーガー、焼き菓子、ドリンク / Silk Burger, baked goods, and drink / 丝绸汉堡，烘焙食品和饮料

まゆふわバーガー
Mayufuwa (cocoon fluffy) Burger
Mayufuwa (茧蓬松) 汉堡

まゆふわ - かぶ
Mayufuwa - Turnip
Mayufuwa· 芜菁

まゆふわ - いちご
Mayufuwa - Strawberry
Mayufuwa - 草莓

高校生がデザインした「シルクリーム」
"Silkcream" designed by high school students
高中生设计的 "Silkcream"

フルーツ エクレア
Fruit Eclair / 水果埃克莱尔

人気のクッキーやラスク
Popular cookies and rusks / 受欢迎的饼干和面包干

鶴岡土産として隠れた人気
Hidden popularity as Tsuruoka souvenir
隐藏的人气鹤冈土特产

ロイヤルミルクティー
Royal milk tea / 皇家奶茶

食材としてのシルクと桑の葉を堪能する オンリーワンなフルコース「庄内シルク物語」
"Shonai Silk Story": the only one full course in the world to enjoy silk and mulberry leaves as food ingredients
世界上唯一以丝绸和桑叶为食品成分的全丝系列料理全套餐【庄内丝绸的故事】

デザートとコーヒー
Dessert and coffee / 甜点和咖啡

クレープは、織りあげた布のイメージ
Crepe is a woven up cloth image
可丽饼是编织布的形象

オードブル「松ヶ丘の想い」
Hors d'oeuvre is "Memory of Matsugaoka"
开胃菜【松冈的回忆】

シルク入りクレープは、もちもちとした食感
Silk crepe will have a chewy texture
真丝填充的可丽饼具有耐嚼的质地

メレンゲのカイコ
Silkworm made of meringue
蛋白酥皮制成的蚕

シルク エクレア
Eclairs with silk
丝绸埃克莱尔

デザートは、その日によっていろいろ
Desserts vary depending on the day
甜点因天而异

桑の葉のピュレの色が美しいスープ
Beautiful soup with mulberry leaf puree's color
桑叶菜泥色的很漂亮汤

魚料理
Fish dishes
鱼菜

肉料理
Meat dish / 肉菜

P.134. 参照 / See page 134 / 请参阅第 134 页

オードブルには、その日の旬の食材のほか、ばんけ（ふきのとう）みそや、
松ヶ岡の郷土料理「いもごぼたもち」など、庄内の味が満載

In addition to the seasonal ingredients of the day, the hors d'oeuvres are packed with Shonai's taste such as BANKE (Fukinoto) soybean paste and Matsugaoka's regional dish "IMOGO rice cake dumpling"
开胃小菜充满了时令食材，还有庄内的风味，例如 BANKE（款冬芽薹）大酱和松冈的当地菜“IMOGO Botamochi”

オードブルのソースは、
松ヶ岡の土から芽吹く桑を表現

Hors d'oeuvre sauce expresses mulberry that sprouted from the soil of Matsugaoka
开胃菜酱体现了从松冈丘陵土壤发芽的桑树

シルクパン - まゆふわ
Silk bread Mayufuwa
真丝面包 -Mayufuwa

「シルクは栄養的にもすぐれた食材、粉ものと合わせて使うと柔らかさやしっとり感が長持ちします」調理長談。

"Silk is a superior cooking ingredient nutritionally, when it is used with flour and, the softness and moisture lasts longer" talk of the cooking chief.

烹饪总监说："食用蚕丝在营养上是优越的，当和面粉一起使用时，其柔软性和水分会持续更长的时间"

鶴岡シルク製ドレス / Wedding dress made of Tsuruoka silk / 鹤冈真丝制成的婚纱礼服

グランド エル・サンには、官民一体の鶴岡市の「鶴岡シルクタウン・プロジェクト」に賛同してつくられた「鶴岡シルク」とのコラボドレスがあります。また、鶴岡土産の「食べるシルク」を、山形県立鶴岡中央高等学校や、山形県立庄内農業高等学校の活動とコラボして、シフォンケーキやシュークリームを販売しています。

At Grand el Sun, there is a dress collaborated with "Tsuruoka Silk", which was founded in favor of "Tsuruoka Silktown Project" in Tsuruoka City, a joint project of public and private sectors. In collaboration with Yamagata Prefectural Tsuruoka Chuo High School and Yamagata Prefectural Shonai Agricultural High School, we sell chiffon cakes and cream puffs as Tsuruoka souvenir "Edible silk".

太阳大爷 (Grand el Sun) 拥有与“鹤冈丝绸公司”(Tsuruoka Silk) 的合作连衣裙，该服装的推出是为了支持官民共办的鹤冈市“鹤冈丝绸城项目”。与山形县立鹤冈中央高中和山形县立庄内农业高中合作，销售丝雪纺蛋糕和丝手指泡芙作为鹤冈的纪念品“食用丝”。

食べるシルクで血液はきれいになるの？

Can edible silk make our blood clean? / 食用蚕丝能清洁我们的血液吗？

セラム - シルクフィブロイン ： ドクターセラム㈱

SERUM-Silk Fibroin: Doctor Serum Co., Ltd. / Serum-Silk Fibroin (丝素蛋白): Doctor Serum Co., Ltd.

2016年4月、事業者の責任において表示する「機能性表示食品」の制度が発足しました。

ドクターセラム㈱では、エビデンス(科学的根拠)に基づいて表示する、健康の維持及び増進に役立つ食品の機能を表示するために、ランダム化比較試験(RCT)に申請し、機能性表示食品として消費者庁長官に届け出をしました。

In April 2016, the system of Foods with Functional Claims was launched and it allows manufacturers to indicate the benefit of functional foods on their own responsibility.

Dr. Serum Co., Ltd. submitted a completed notification and related documents for Serum-Silk Fibroin as Foods with Function Claims to the Secretary-General of the Consumer Affairs Agency, based on data of the randomized controlled trial (RCT).

2016年4月，启动了【注明功能的食品】系统，该系统使制造商能够自行怀着责任感功能性食品的好处。Serum Co.,Ltd. 根据随机对照试验(RCT)的数据，向消费者事务管理局秘书长提交了一份完整的通知书和有关功能性食品的Serum-Silk Fibroin的相关文件。

セラム - シルクフィブロイン
Serum-Silk Fibroin / Serum-Silk Fibroin

主原料：シルクフィブロイン
Main ingredient: Silk fibroin / 主要成分: 丝素蛋白

ゼリー状（1包10g）
Jelly (10 g / pack) / 果冻形状(每包10克)

ピーチ果汁の味と色
Flavor and color of peach juice / 桃汁的味道和颜色

食べた場合の機能性 / Functionality when eaten / 食用后的功能

セラム - シルクフィブロインは、1,000kgの繭から160kgだけ精製できるフィブロインです。体内では分解されず、腸を通過しながらナノレベルの微細構造の中に、油分と糖分を吸着して体外に排泄する機能が期待されます。

近年LDLコレステロール値が170mg/dLを超えていた著者の中山は、2016年6月からセラム-シルクフィブロインの治験に参加し、2か月後に168 mg/dL、3か月後には154mg/dL、8か月後の2017年1月に121mg/dLになりました*。

この後５ページにわたって、「体験者の声」を紹介します。

Serum-Silk Fibroin is the fibroin that can be obtained by only 160 kg after purification of 1,000kg cocoon. This protein is not decomposed in the body and is expected to have the function of adsorbing oil and sugar in the nano-level microstructure while passing through the intestine and excreting it outside the body.

Author Nakayama whose LDL cholesterol level exceeded 170 mg /dl in recent years participated in the clinical trial of Serum silk fibroin from June 2016, 168 mg /dl (two months later), 154 mg /dl (three months later) and 121 mg /dl in January, 2017 (eight months later*).

In the next five pages, we will introduce the "voice of the experience-maker".

Serum-Silk Fibroin 是一从1000公斤茧中纯化后僅能获得160公斤的丝素。这种蛋白质不会在体中分解，穿过肠道時具有在纳米级精细结构中吸附油分和糖分的功能，并将其排出体外。

近年来LDL胆固醇水平超过170mg/dL的作者的Nakayama参加了Serum-Silk Fibroin的临床试验，从2016年6月开始，两个月后的为168 mg/dL，三个月后为154mg/dL。在8个月后的2017年1月为121 mg/dL*。

在接下来的五页中，我们将介绍“体验者的声音”。

＊中山の治験結果は、掛かりつけ病院などの血液検査の結果を転載したものです。

＊Nakayama's clinical trial results are reprints from the blood test results of the family hospital, etc.

＊中山临床试验结果是家庭医院和其他医院的血液检查结果的转载。

1日1回、1包摂取 / Take 1 pack once a day /每天喝一包一次

マッサージサロンの経営者、女性(58歳)*
Massage salon manager, female (58 years old)*
按摩沙龙総经理，女性 (58 岁) *

体調の推移 / Transition of physical condition / 身体状况的演变

発症：50代、中年期から便秘がちになったり、体重が増えたりするようになる。

開始：2016年3月から1日1包とる。

改善：食事の量を減らさなくても体重が維持できるようになる。血圧や血糖値、中性脂肪の数値も異常なし。

現在：50代、ダイエットのストレスがなくなり、身体のキレを取り戻す。

Symptom (applicant of 50s): I was apt to be constipated and gain weight from middle age.
发病：自从50多岁中年以来，便秘趋于增加，体重也开始增加。

Start: I took 1 pack per day from March 2016.
开始：从 2016 年3月开始每天喝1小包。

Improvement: I got to keep my body weight constant without reducing diet. No abnormality in blood pressure, blood sugar level, and triglyceride level.
改善：可以維持体重而不減少饮食，而血压，血糖水平和甘油三酸酯水平也很正常。

Present situation: No stress for diet, and body's sharpness is recovered.
当前：饮食压力消失了，我恢复了身体的清正常指标。

50代から便秘がちになり、身体も重く感じられ、年齢を実感するようになりました。

そのころ、45歳の友人から「食事制限をせずに、らくにダイエットができる方法はないでしょうか」と、相談を受けました。私も今後、同じ悩みを持ちそうだと感じ、ちょうどいい機会と思い、来店するお客様たちから情報を集めたところ、絹のゼリーの存在を知りました。

絹のゼリーは、たんぱく質の微細な穴に油や糖を吸着し、脂っこい料理のほか、ご飯やめん類が好きな人にぴったりの健康食品とのこと。早速絹のゼリーをとり寄せ、私、質問者、さらに別の友人の計3人で、朝・昼・晩の中で、最も食事量の多いときの食前に飲み始めました。

絹のゼリーは、フルーツゼリーのようで飲みやすく水も不要、長期間飲み続けられると思いました。

私は、ただ1包飲むだけなのに数日後にはお通じがよくなったことに気づき、いつの間にか毎日便通があるようになり、体重変動はありませんでしたが、身体が軽くなったように感じました。

From 50s, I was apt to be constipated and felt my body heavy, and began to recognize my physical condition corresponding to my actual age. At that time, a 45-year-old friend asked me, "Don' t you know an easy diet method without diet restriction?" I felt that I might have similar problem in future and I thought this was a good opportunity to find the solution of this problem. And I gathered information from clients who came to my store and I knew the existence of silk jelly.

Silk jelly is explained to be a healthy food that is very favorable to people who like rice and noodles as well as greasy foods because it adsorbs oil and sugar in the fine holes of protein. So I readily ordered silk jelly, and 3 persons (I, my friend who inquired me and another friend) began taking it before a meal (one time a day before the heaviest meal among breakfast, lunch or dinner).
I felt silk jelly is easy to drink like fruit jelly and no water is needed. I felt I could continue taking it for a long time.
Although I took only one pack a day, a few days later I noticed that my bowel movement became better, and without realizing, I got to have a bowel movements every day. Though there was no body weight change, I felt as if my body got lighter.

自从我50多岁以来，我便开始便秘，身体沉重，开始感到自己真是上了年纪。大约在那个时候，一个 45 岁的朋友问我：您知道没有节食的简单饮食吗?。我也觉得我可能有类似的问题，我认为这是一个机会，当我从拜访该商店的顾客那里收集信息时，我就知道了丝果冻的存在。

丝果冻的在蛋白质中的微小孔中吸附油和糖，是一种健康食品，非常适合喜欢米饭，面条和油腻菜肴的人们。因此，立即订购了丝果冻，总共有3个人，包括想咨询的一位朋友和另一位朋友和我，在早上，下午晚上用的餐最多的时候就开始饭前喝。
我以为，丝冻像果冻一样易于饮用，不需要水，我以为我可以继续喝很长时间。尽管我每天只喝一包，但几天后，我注意到我的排便情况变得很好，并且每天都有排便的经历，我没有减轻体重，但我却感到自己的身体更轻。

＊引用文献：『健康365』2017年6月号 P. 80-81。年齢は、出版当時。

＊References: "Health 365" June 2017, P. 80-81. Age was at the time of publication.

＊参考:《健康365》2017年6月号 第80-81页。 年龄是发布时的时间。

友人たちの体調の推移 / Transition of physical condition of my friends / 朋友身体状况的过渡

45 歳の友人は体重が増え続けていて、メタボリックシンドローム(内臓脂肪症候群)になる不安を抱えていました。絹のゼリーをとり始めてからは運動もしていないのに、「食事の量を減らさずに、体重の増加が止まった」のだそうです。

そこで、2 人とも血液検査をしたところ、血圧や血糖値、中性脂肪の数値に異常はみられませんでした。もう 1 人の友人は、飲酒前に 1 包とると翌日二日酔いをしないとのことです。

45-year-old friend was constantly gaining weight and was worried about developing metabolic syndrome. He said that his body weight increase stopped without dieting since he had started to take silk jelly, though he had made no Special exercise. So my friend and I received blood test and no abnormality was found for both in blood pressure, blood sugar, and triglyceride levels. Another friend said that take one pack before drinking alcohol, that not get a hangover the next day.

45岁的朋友不断体重增长，并担心会发展为代谢综合征。他说，尽管自从开始服用丝凉粉以来他就一直没有运动，但他说："体重的而又不减少饮食我已经停止增加"。因此，我们俩都进行血液检查时，血压，血糖水平和甘油三酸酯水平均未发现异常。另一个朋友说喝酒前服用1小包，第二天不要宿醉。

1日2回、各1包摂取 / Take 1 pack each, twice a day / 每天喝两次，每次一包

食品関連会社勤務、女性(42歳)*
Worked at a food-related company, female (42 years old)* / 在一家食品相关公司工作，女性(42岁)*

体調の推移 / Transition of physical condition / 身体状况的演变

発症：22歳、入社後の健康診断でたんぱく尿が出る。精密検査で膠原病と診断される。

悪化：20代、全身性エリテマトーデスによる炎症で腎機能低下。たんぱく尿 3+、総コレステロール値が 220mgを超える。

開始：42歳、2016 年9月から、昼・晩の食前に1包ずつとる。

改善：1ヵ月後たんぱく尿が出なくなり、総コレステロール値は190mg となる。

現在：疲れにくくなり、体力が向上し、旅行を満喫している。

Symptom: 22 years old, protein in urine was detected at the medical examination after joining the company. After a detailed examination, diagnosed with collagen disease.
发病：22岁，加入公司后，身体检查时出现蛋白尿。经过仔细检查，诊断为胶原病。

Exacerbation: 20s, renal function drop due to the inflammation by systemic lupus erythematosus. Proteinuria 3+, total cholesterol level exceeds 220mg.
恶化: 20 多岁时，由于系统性红斑狼疮引起的炎症导致肾功能下降。蛋白尿 3+，总胆固醇水平超过 220mg。

Start: 42 years old, since September 2016, I have taken 1 pack before lunch and dinner.
开始：42岁，自2016年9月起，在午餐和晚餐前吃了1小包。

Improvement: One month after ingesting silk jelly, proteinuria was not found and total cholesterol level became 190 mg.
改善: 摄入蚕丝后一个月，蛋白质尿液消失，总胆固醇水平变为190mg。

Present situation: don't get tired easily, improved body strength, enjoying trips. / 当前：减轻了疲劳，提高了体力，并且喜欢旅行。

子どもの頃から、アトピー性皮膚炎 (以下アトピー) がひどく、22 歳で働き始めてから常に疲労感に悩まされてきました。

入社後に受けた健康診断で蛋白尿が出て、精密検査の結果、全身性エリトマトーデスという膠原病への罹患がわかりました。それからの20年間はステロイドの服用と、毎月の血液検査が欠かせませんでした。

Atopic dermatitis (hereinafter called atopy) has been serious since childhood, and I have always been plagued by the feeling of exhaustion since I started working at age 22. Proteinuria was detected at a medical examination after joining the company, and through a detailed examination, I was found affected with connective tissue disease (a systemic lupus erythematosus).
For the following 20 years, taking steroids and monthly blood tests were essential.

自从儿童期开始，特应性皮炎(以下称为特应性)就很严重，自22岁开始工作以来，我总是很累。加入公司后，在体检中检测到蛋白尿，精确检查发现，诊断为胶原病，系统性红斑狼疮。在接下来的20年中，服用类固醇，并每月进行血液检查。

*引用文献：『健康 365』2017 年 4 月号。P. 72-73。年齢は、出版当時。

*References: "Health 365" April 2017, P. 72-73. Age was the time of publication.

*参考:《健康365》2017年4月号。第72-73页。年龄是发布时的时间。

私がかかった膠原病は、全身性エリトマトーデス（SLE）という、全身の様々な臓器に炎症が起こるタイプで、特に女性の発症が多いといわれているそうです。アトピーがひどくて疲れやすいのは、膠原病が原因、そして蛋白尿は膠原病の合併症として起こった腎炎によるものでした。

検査結果が悪いときには2ヶ月半もの入院加療となって、業務に支障が出るのが気になりました。

毎月の血液検査の結果がよいようにと、減塩、低タンパクの食事を心がけていても、40代になるまで、蛋白尿は常に1+ ～ 2+を推移し、コレステロール値は常に基準値を超えていました。

検査結果の改善の兆しもない中、長期入院による業務への支障という可能性にめげていた私に、2016年8月末、同僚が絹のゼリーをすすめてくれました。

絹の主成分であるフィブロインが、体内の余分な脂肪を吸着して排泄することで、高いコレステロール値や血糖値を下げる働きがあるとのことです。私の体質に合うかどうかは不明でしたが、すぐに試してみたいと思いました。

９月から試して、９月の末の検査時には総コレステロール値が190mgまで下がりました。また、８月の検査で1+だった蛋白尿がでませんでした。

それに、予想外の嬉しい効果がありました。４月から取り組んでいた、野菜中心の食事で腹八分にするダイエットで、８月までに5kg減量しましたが、それ以上は減らなかったのです。しかし、絹のゼリーを食べてからはするすると落ちて、10月までに9kg減りました。

これまでは疲れるのが嫌で、旅行に誘われても断ることが多かったのですが、絹のゼリーで体力に自信が持てるようになって、オーストリア旅行を楽しむことができました。

もう少ししたらジムに通って、もっと体力をつけたいと思っています。

Connective tissue disease that I suffered from is called systemic lupus erythematosus (SLE), a type in which inflammation occurs in various organs of the whole body, and its onset is said to be more frequent in women. It became clear that severe atopy and getting tired easily was caused by connective tissue disease and proteinuria was caused by nephritis that occurred as a complication of connective tissue disease. I was worried about the test result, because, in case of unfavorable test results, I would have to be hospitalized for two and a half months, and my work would be hindered. With the hope of favorable monthly blood test result, I continued a low-salt and low-protein diet, however my proteinuria was always between 1+ and 2+, and cholesterol levels exceeded the standard value until I became 40 years old.

Under the situation that I continued to worry about the possibility of hospitalization owing to no improvement at my test result, my colleague recommended me to take silk jelly at the end of Aug., 2016. It was explained that silk fibroin, the main ingredient of silk, has the function of lowering high cholesterol and blood glucose levels by adsorbing and excreting extra fat in the body. I didn't know if it would suit my constitution or not, but I wanted to try it immediately.

I tried it from the beginning of September and my total cholesterol level dropped to 190mg at the end of September examination. In addition, proteinuria which was 1+ at August inspection was not detected. And it had an unexpected and pleasing effect. With the vegetable-based diet that I had been working on since April, I lost 5kg by August, but I couldn't lose more. However, after eating the silk jelly, the body weight decreased steadily, and by October I lost 9 kg.

So far, I often refused invitation to travel, because I disliked to get tired. But I became confident in my physical strength with silk jelly, and could enjoy traveling to Austria.
In near future, I would like to go to the gym and improve my physical strength.

我遭受的胶原蛋白疾病是系统性红斑狼疮（SLE），这是一种发炎现象，发生在全身的各个器官中，据说尤其发生有很多女性案例。特应性疾病，严重且容易疲倦由胶原蛋白病引起的，和蛋白尿是由于肾炎引起的，它是胶原蛋白疾病的并发症。当测试结果不好时，我住院了两个半月，我担心自己的工作会受到阻碍。我吃了低盐低蛋白饮食，以改善每月的血液检查结果，然而，但是直到40多岁，我的蛋白尿一直保持在1+和2+之间，并且我的胆固醇水平超过了标准值。

当测试结果没有改善的迹象，但由于长期住院可能会影响我的工作，2016年8月底，一位同事向我推荐了丝果冻。据说丝的主要成分丝素具有通过吸收和排泄体内多余的脂肪来降低高胆固醇和血糖水平的功能。我不知道它是否适合我的身体，但我想马上尝试一下。

我从9月初开始试用，9月的底测试时我的总胆固醇水平降至190mg。此外，在未检测到蛋白尿（在8月的测试中为1+）。产生了意想不到的令人愉悦的效果。自4月以来我一直在努力的饮食是通过吃蔬菜为主的饮食来减轻体重，到8月我减掉了5公斤，它没有进一步减少。但是，吃了丝凉粉后，体重稳步下降，到10月，体重下降了9公斤。

直到现在，我都不想累，所以当我被邀请去旅行时，我经常拒绝，但是丝绸果冻让我对自己的体力更有信心，并且能够在奥地利旅行。
从现在开始，我想去健身房锻炼身体体力。

1日3回、各1包摂取 / Take 1 pack each 3 times a day / 每天喝三次，每次一包

貿易関連会社勤務、男性（45 歳）*
Trade related company work, male (45 years old)* / 与贸易相关的公司工作，男性（45 岁）*

体調の推移 / Transition of physical condition / 身体状况的演变

発症：44歳、健康診断で高血糖を指摘されるも、1年間放置する。

悪化：45歳、健康診断で空腹時血糖値が125mgとわかり、糖尿病の1歩手前と指摘される。

開始：悪化対策として、2014年4月から、朝･昼･晩の食前に1包ずつ摂取。

改善：飲用1ヵ月後に身体が軽くなり、3ヵ月後には体重が2kg減って、おなかまわりがスッキリ。

現在：45歳、身長 170cm 体重 61kg、体調不安がスッキリ。

Symptom: 44 years old, high blood sugar was pointed out by medical examination, but left untreated for 1 year.

发病：44 岁，经身体检查指出血糖高，然而我忽略一年。

Exacerbation: 45 years old, a medical examination revealed a fasting blood glucose level of 125 mg, pointed out that it is one step before diabetes.

恶化：在 45 岁时，身体检查显示空腹血糖水平为 125 mg，指出的是离糖尿病只差一个步。

Start: As a measure against exacerbation, I started to take one pack before each meal in the morning, noon and evening from April 2014.

开始：为了预防病情恶化，我从2014年4月起在饭前，早，午，晚各我喝了一包。

Improvement: The body became lighter after 1 month of ingestion and body weight decreased by 2 kg after 3 months of ingestion, and the periphery of abdomen get slimmer.

改善：喝1个月后身体变轻，3个月后体重减轻了2千克，肚子周围感到轻松。

Present situation: 45 years old, height 170 cm, weight 61 kg, anxiety to physical condition disappeared.

当前：45 岁，身高 170 厘米，体重 61 公斤，身体状况令人耳目一新。

若いころから食べることが大好きで大食いで早食いでした。社会人になってからは毎日が忙しく、さらに早食いになりました。食生活の異常に気がついたころ会社の健康診断で高血糖と指摘され、定期的に運動をすることと、糖質と油を控えることを指導されたものの、1年間放置していました。

糖尿病にならないために何ができるのか悩む中で、仲のよい友人に糖質と油を排出する絹のゼリーを紹介されました。

絹のゼリーを朝･昼･晩の食前に1包ずつとった1ヵ月後には身体が軽くなったように感じられ、3ヵ月後には体重が2kg減りました。

From my younger age, I love to eat, in addition, big volume at fast speed. Since I became a member of society, I've been busy every day, and I've eaten even faster. When I noticed an abnormality in my eating habits, I was pointed that my blood sugar was high at a company medical examination. I was instructed to exercise regularly and to refrain from carbohydrates and oil, but I took no action for a year.

While I was worried about what I could do to prevent diabetes, my intimate friend introduced me a silk jelly that excretes sugar and oil. I felt that my body became lighter one month after I took one pack of silk jelly before each meal of breakfast, lunch and dinner, and after 3 months I lost 2 kg.

我从小就喜欢吃东西，所以我吃了很多，又吃得很快。成为社会成员之后，我每天都很忙，而且进食速度甚至更快。当我发现自己的饮食习惯出现异常时，公司的医疗检查表明我的血糖很高，并被指示定期运动并避免碳水化合物和油脂，且这一年来没有改变过。

当我担心如何预防糖尿病时，一个好朋友向我介绍了一种丝果冻，它可以排出糖和油。我在早餐，午餐和晚餐前吃了一包丝冻后一个月，我的身体变轻了，三个月后，我减了2公斤。

バラ園経営、女性（59 歳）**：体調の推移
Rose garden management, female (59 years old)**: Transition of physical condition / 玫瑰园管理，女性（59 岁）**：身体状况的演变

発症：55歳、蜂窩織炎によるストレスで、空腹時血糖値が300mgを超えて、糖尿病と診断される。

Symptom: 55 years old, By the stress due to cellulitis, fasting blood glucose exceeds 300 mg, diagnosed as diabetes.

发病：55 岁，由于蜂窝织炎引起的压力，空腹血糖水平超过 300 毫克，并诊断出患有糖尿病。

*引用文献：『健康 365』2014 年 10 月号。P. 59-60。年齢は、出版当時。
**引用文献：『健康 365』2016 年 12 月号。P. 73-74。年齢は、出版当時。

*References: "Health 365" October 2014, P. 59-60. Age was at the time of publication.
**References: "Health 365" December 2016, P. 73-74. Age was at the time of publication.

*参考：《健康365》2014年10月。第59-60页。年龄是发布时的时间。
**参考：《健康365》2016年12月。第73-74页。年龄是发布时的时间。

開始：55歳、2012年12月から、朝·昼·晩の食前に1包ずつとる。

改善：56歳、1年後、300mgを超えていた空腹時血糖値が160mgに、12.8だったHbA1cが6.5になる。体重が、88kgから77kgに減少。

現在：59歳、絹のゼリーをとって、体調を維持しながら、バラの栽培に取り組んでいる。

Start: 55 years old, I took one pack before each meal of breakfast, lunch and dinner from December 2012.
开始：55岁，我从2012年12月开始在早餐，午餐和晚餐前喝了1小包。

Improvement: 56 years old, one year later, the fasting blood glucose level, which was over 300 mg, was 160 mg, and HbA1c was also lowered from 12.8 to 6.5. Weight decreased from 88 kg to 77 kg.
改善：56岁，一年后，空腹血糖水平从300 mg减了到160 mg，HbA1c从12.8减了到6.5。重量从88公斤减少到77公斤。

Present: 59 years old, while taking silk jelly and maintaining physical condition, I am engaged in cultivating roses.
当前：59岁，服用丝果冻并正在耕种玫瑰，同时保持身体健康。

知人に紹介されたマッサージ院で聞いた「余分な脂肪を吸着して排出する絹のゼリーの働き」に感心して、1日3回とりいれたことで、体重が11kgも落ちました。

I was impressed with "the function of silk jelly adsorbing and discharging extra fat" which I heard at a massage institution introduced by an acquaintance, and so I took it three times a day. I lost 11kg.

在一家熟人介绍的按摩机构中听到的“吸收和排出多余脂肪的丝冻的功能”给我留下了深刻的印象，所以我每天吃三遍，我瘦了11公斤。

1日4回、各1包摂取 / Take 1 pack each 4 times a day / 每天喝四次，每次一包

女性（68歳）*：体調の推移 / Female (68 years old)*: Transition of physical condition / 女性（68岁）*：身体状况的演变

発症：63歳、倦怠感やめまいを覚えるようになる。

悪化：63歳、HbA1cは11.4、空腹時血糖値が400mg以上と、重度の糖尿病と判明。さらに、糖尿病腎症を合併していることもわかる。

開始：63歳、2013年から、朝·昼·晩の食前と就寝前に1包ずつとる。

改善：飲用3ヵ月後、HbA1cが7.4、空腹時血糖値が240mg前後になりました。

現在：68歳、めまいやだるさが解消、腎機能の悪化もありませんでした。

Symptom: 63 years old, I begins feeling tired and dizzy.
发病：63岁，我开始感到疲倦和头晕。

Exacerbation: 63 years old, HbA1c was 11.4, fasting blood glucose level was 400 mg or more, and found to be severe diabetes. Further, diabetic nephropathy was found coexisting.
恶化：63岁，HbA1c为11.4，空腹血糖为400mg或更高，被发现是严重的糖尿病。此外，它还表明糖尿病性肾病被合并。

Start: 63 years old, from 2013, I took 1 pack before each meal of breakfast, lunch, and dinner, and before bedtime.
开始：63岁，它是从2013年开始，我在饭前早餐，午餐，晚餐之前和睡時服用1小包。

Improvement: After taking 3 months, HbA1c became 7.4 and fasting blood glucose level became around 240 mg.
改善：3个月后，HbA1c为7.4，空腹血糖水平约为240mg。

Present situation: 68 years old, dizziness and lassitude disappeared, renal function did not deteriorate.
当前：68岁，头晕乏力消失，肾功能未恶化。

私は20年前に腎臓結石の手術で入院したことがあり、その時に病院で受けた治療にたいして強い不信感があります。それで、今回医師に検討することを告げられた人工透析はどうしても避けたいと思いました。

友人にすすめられた絹のゼリーが、血液内の余分な糖と油を排泄し、腎臓の働きを助ける可能性があることに、掛ける気になったのです。

I was hospitalized for kidney stone surgery 20 years ago, and I have a strong distrust against the treatment I received in the hospital at that time. So, this time when my doctor told me to consider artificial dialysis, I really wanted to avoid it. I wanted to bet silk jelly recommended by a friend because it excretes extra sugar and oil from the blood and could help the kidneys work.

20年前，我曾因肾结石手术住院，对当时的治疗方法我非常不信任。所以，这次我的医生告诉我要考虑的是人工透析，但是我真的想避免这种情况。我被朋友推荐的丝果冻可以排泄血液中过量的糖和油并帮助肾脏工作的事实激发了我的动力。

*引用文献：『健康365』2018年4月号。P. 56-57。年齢は、出版当時。

*References: "Health 365" April 2018, P. 56-57. Age was at the time of publication.

*参考：《健康365》2018年4月号。第56-57页。年龄是发布时的时间。

世界初！ウスタビガ大量飼育
World's first! Rhodinia fugax mass rearing / 世界第一！ Ustabiga大规模饲养

富士発條㈱の試み / Attempt of Fujihatsujyo Co.,Ltd. / Fujihatsujyo Co.,Ltd. 的尝试

大量飼育場*(栃木県那須塩原市)
Mass breeding ground* (Nasushiobara, Tochigi) / 大规模繁殖场＊（那须盐原，枥木县）

ヤママユガ科の中でウスタビガは、これまで繭や絹糸腺の利用がなされてきませんでした。

不思議な形を尊んでお守りとして使う地域はありましたが、美しい緑色にひかれて、繭を集めても糸は作れませんでした。

2018年9月、信州大学で開催された日本野蚕学会の展示発表で、サナギを材料にしたサプリメントに出会い、短期間での大量飼育の成功と、製品化にとても驚きました。

クスサンの大量発生を聞くことはありますが、ウスタビガは聞いたことがありません。群れては暮らさない種のようなので、卵や成虫を集めるのは大変だったでしょう。

Up to now, in case of the Usutabiga (Saturniidae family) its cocoon or silk gland has not been used. Although there were areas where the mysterious shape was respected and used as a talisman, thread could not be made from the collected cocoons even if attracted by their beautiful green color.

September 2018, at the exhibition of the Japanese Society for Wild Silkmoth held at Shinshu University, I met a supplement made from pupa and was very surprised by the success of mass breeding in a short period and its commercialization.

We have heard about the mass generation of Kususan, but has not heard that of Usutabiga. Since that species seems not to live in groups, it would have been difficult to collect eggs and adults.

迄今为止，在Saturniidae中，Rhodinia fuga(Usutabiga)从未使用过茧或丝腺。尽管有些地方尊重神秘形状并将其用作护身符，但是美丽的绿色吸引了我，即使我收集了茧也无法制造丝纱了。

2018 年 9 月，在信州大学举行的日本野蚕学会展览会上，我遇到了由蛹制成补充品，我对短期内大规模飼養和产品化的成功感到非常惊讶。我们听说过孳生Kususan的大规模发生，但Usutabiga尚未听说过它。

是由于它不是成群生活的物种，因此收集卵和成虫这将是很多。

孵化幼虫 / Hatching larva / 孵化的幼虫

2〜3齢幼虫 / 2nd to 3rd instar larva / 二至三龄幼虫

天蚕の繭の色素とは異なる、青みの強い
鮮やかな緑色は退色しにくい

Unlike the pigment of the cocoon of the Tensan, vivid green with a strong bluish color does not easily fade.
与天蚕茧的色素不同，明亮的绿色，带有强烈的蓝色调不会轻易褪色。

5齢幼虫 / 5th instar larva / 五龄幼虫

成虫の交尾 / Mating of adults / 成虫交配

サナギで栄養補助食品を作り繭の非繊維利用を計画中
Planning to make nutritional supplements from pupa and to use cocoon other than fiber
用蛹制作营养补品，并计划采用非纤维茧

2015年4月、岩手大学と「野蚕の機能性解明と機能性に着目した製品開発」をテーマに共同研究を開始。翌年2月飼育棟を10棟設営、2017年8月には、繭の生産は7000コに達し、毎年安定的に集繭できるようになったとのことです。

ウスタビガの繭のシルクタンパク質は、家蚕やほかの野蚕と同様に人の皮膚に近いアミノ酸組成で、紫外線を反射吸収し、さらに抗菌性など様々な機能を持つことがわかってきています。現在㈱バイオコクーン研究所（鈴木幸一所長）と共同研究し、ウスタビガのサナギが持つ抗酸化を始めとした様々な機能を活用し、化粧品を含めた今までにない商品開発に取り組んでいます。

サナギを使用したサプリメント
Supplements using pupa
蛹制成的产品

In April 2015, collaborative research with Iwate University started on the theme of "Elucidation of functionality of wild silkmoth and product development focused on its functionality". In the following February, 10 breeding houses were set up, and in August 2017, cocoon production reached 7000 pieces, which means that it became possible to stably gather cocoons every year. It has been known that the silk protein of the cocoon of Usutabiga, like that of domesticated silkworm and other wild silkworm, has an amino acid sequence close to that of human skin, reflect and / or absorb ultraviolet ray, and has other various functions such as antibacterial properties. Currently, we collaborate with Biococoon Research Laboratories (Director Dr. Koichi Suzuki) and by utilizing various functions such as antioxidation possessed by Usutabiga pupa, we are working on unprecedented product development including cosmetics.

2015年4月，我们与岩手大学开始了以“阐明蚕的功能和注重功能的产品开发”为主题的联合研究。在接下来的2月中，建立了10座建筑物饲养室，到2017年8月，茧的产量达到7000枚，这意味着每年都有可能稳定地收集茧。众所周知，Usutabiga 茧的蚕丝蛋白像驯养的蚕和其他野生蚕一样，能够反射和吸收紫外线并具有多种功能，例如抗菌性能。目前，我们与 Biococoon Research Laboratories（导演铃木幸一博士）合作，并利用Usutabiga蛹拥有的多种功能（例如抗氧化作用），我们致力于包括化妆品在内的前所未有的产品开发。

＊66-67ページの画像提供は富士発條㈱
http://fujihatsujyo.jp/bio/index.html

＊Image on page 66-67 is provided by Fujihatsujyo Co.,Ltd.
http://fujihatsujyo.jp/bio/index.html

＊第66-67页的图像由Fujihatsujyo Co., Ltd. 提供。
http://fujihatsujyo.jp/bio/index.html

未来食として絹糸昆虫食をひろげる活動

Activities to spread silk insect usage for future food / 推广蚕丝昆虫食品，作为未来食品的活动

ワイルドシルクミュージアム出張講座
Visiting lecture at WILD SILK MUSEUM / 商务旅行讲座野生丝绸博物馆的

養蚕関係者と旬を楽しむパーティー
Party to enjoy season with people involved in sericulture / 与养蚕人员一起享受季节性聚会

サナギになりたての「旬のおカイコを食べる会」が安中市で開催され、エリサンのサナギペーストや蚕糸昆虫の糞茶も試食・試飲しました。

The "seasonal silkworm eating party" with silkworm that just turned into pupa was held in Annaka City, and we sampled Eri silkworm pupa's paste and silk insects poo tea.

刚刚变成蛹后，【季节性蚕食聚会】在安中市举行，我们还在那里試吃試飲了Erisilkworm蛹的糊，和吐丝线昆虫粪茶。

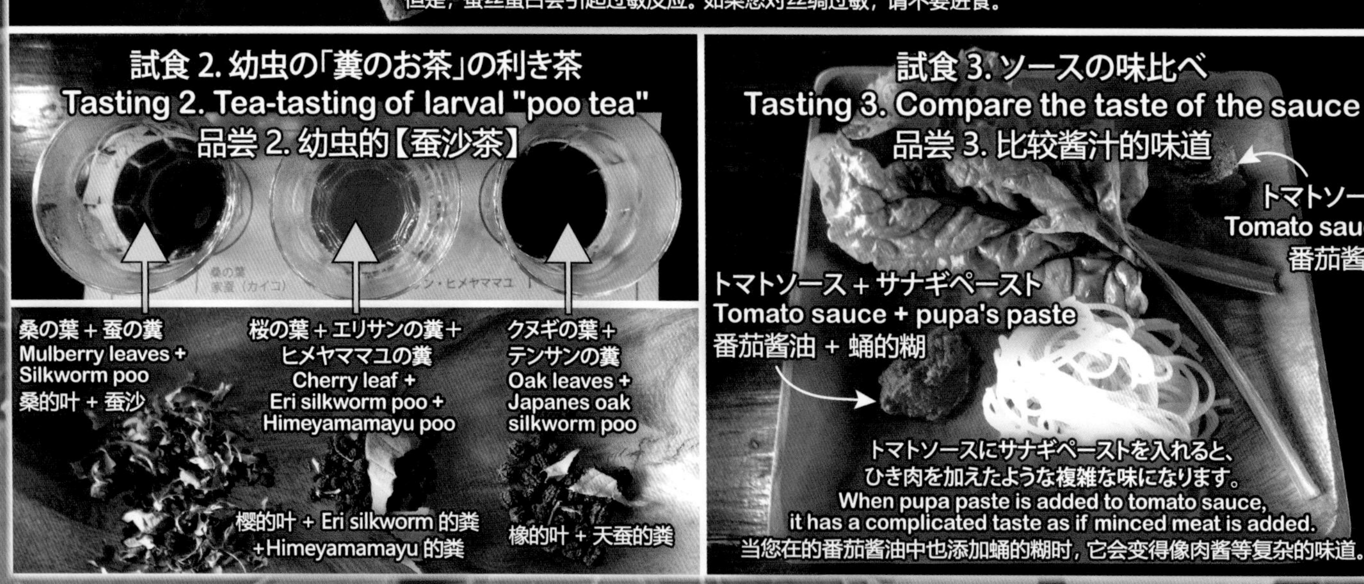

試食 4. 旬のカイコを食べる
Tasting 4. Eat seasonal silkworms / 品尝 4. 在食用蚕的最佳季节

この試食会は、群馬県安中市の「ton-cara」で開催されました。https://www.ton-cara.com
This tasting party was held at "ton-cara" in Annaka, Gumma.
这次品尝会，在群马县安中市的 " ton-cara" 举行。

手工芸品の材料 / Handicraft's materials / 手工艺品的材料

カンボジアの黄色いシルクの「キビソ糸」で製作したランプシェード＊（カンボジア、IKTT 伝統の森にて）
Lamp shade * made with Cambodian yellow silk "*Kibiso* thread" (Cambodia, At the IKTT Traditional Forest)
柬埔寨黄色真丝 Kibiso 线制成的灯罩＊(柬埔寨, 在的 IKTT 传统森林)

＊ランプシェード：糊に浸した糸をゴム風船に巻き、乾燥後に風船をとり除きます（製作&撮影=増間 扶佐子氏）。関連作品はワイルドシルクミュージアムに常設展示されています。

＊Lampshade: Winding threads soaked in glue around a rubber balloon and removing it after drying (Production & shooting = Ms. MASUMA Fusako). Related works are permanently displayed at the WILD SILK MUSEUM.

＊灯罩：将浸有胶水的用线包裹橡胶气球中, 干燥后取下气球 (制造与摄影=増間扶佐子先生)。相关作品永久陈列在野生丝绸博物馆。

保湿&放湿性・防臭性に優れ、肌に優しい
Excellent in moisture retention & its release and deodorization and gentle to the skin
极佳的保湿效果，释放水分，除臭剂，对皮肤温和

シルクは、人の肌に近いアミノ酸で構成されているために快適な下着になります。

薄く軽く、乾燥する季節にも肌をしっとり保ち、汗をかいても吸収して放散する特性を持ち、繊維に湿気をため込まず、べたべたと肌につかず、汗で湿った肌着で冷える心配がありません。

シルクは保温力がありながら暑くはならず、寒暖差に対しての調節がきき、美肌効果も期待され、肌着に向いた繊維です。

夏は生糸のニットがサラッとして心地よく、冬は絹紡糸や真綿糸のニットが温もりがあり、重ね着せずに、暖かく過ごせます。

一方、シルクのアンモニア消臭率は 99% と高く(P.106)、介護関連素材などにも向いています。

近年では、シルクの保湿&放湿性・保温性・防臭性・UVカットなどの特性(P.58)を利用した、新しい商品展開の試みが各所で始まっています。

Silk can be a comfortable underwear material because it is composed of amino acids close to human skin.

Underwear made of silk is thin, light and keeps the skin moist even in dry season. Silk has excellent thermal insulation capability, so body temperature can be kept within comfortable range against warm and cold temperatures don't have to worry about getting cold with sweaty and skin beautifying effect can be expected and so it is the fiber very suited to underwear.
In summer, raw silk yarn is smooth and comfortable, and in winter, spun silk yarn and floss silk yarn are warm and can stay warm without layering.

On the other hand, the ammonia deodorization rate of silk is as high as 99% (P.106), and it is suitable for products such as nursing care related materials. In recent years, attempts to develop new products are beginning in various places, using silk's properties such as moisture retention & desorption, heat retention, deodorization and UV protection(P.58).

丝绸是它将是舒适的内衣，因为它由接近人体使的氨基酸组成。

薄而轻。即使在干燥季节也能保持皮肤湿润，甚至在出汗时也会吸收和通气，它不会在纤维中积聚水分，因此它不会粘在皮肤上并且被汗水弄湿，而使皮肤感。不用担心汗湿的内裤会着凉。丝绸具有保温性，但感到可调节闷热和温差，适用于内衣的纤维，且具有美容皮肤的效果。
在夏季，生丝纱光滑而舒适，而在冬季，绢丝纱和丝绵线纱是温暖的，可以保持温暖且不会没有分层。

另一方面，丝绸的氨气除臭率高达99％(第106页)，适用于做护理品的相关材料。近年来，在各个地方开始尝试利用丝绸的保湿和水分释放特性，保温特性，除臭特性，与諸如特性(第58页)防紫外线特性来开发新产品。

＊上記左右の画像提供：㈱シルクマルベリー P.134

*Provide left and right images: Silk Mulberry Co., Ltd. P.134

＊提供左右图像: Silk Mulberry株式会社 P.134

日常着としてのシルク / Silk as everyday wear / 丝绸作为日常穿着

インドや東南アジアの野蚕の布には、非常に毛羽立つ布があります。

着用後毛羽立つので、毛羽立たない布に慣れた日本の多くの消費者には馴染みにくく、購入先にクレームをつけます。

しかしそれらの布は、ゆるいよりの糸で織られ、柔らかい質感と暖かさを追求した布であって、こすれによる毛羽立ちは問題視されない織物です。

Within wild silk clothe in India and Southeast Asia, there is a very fuzzy cloth. Because it becomes fuzzy after wearing, it is difficult to use it for many Japanese consumers who are accustomed to well-prepared cloth, and sometimes consumers receive complaints.

However, these fabrics, being woven with loose twisted threads, pursue soft texture and warmth, and so fluff caused by rubbing is not regarded as a problem.

来自印度和东南亚的一些野蚕丝布，起毛起球现象。由于穿着后会变得起毛，对于许多习惯于使用不起毛布料的日本消费者来说，这很困难，并且会引起的抱怨。

然而，这些布是用松散的绞线编织的，并且追求柔软的质地和保暖性而编织，因摩擦引起的起毛，不是問題可以忽略。

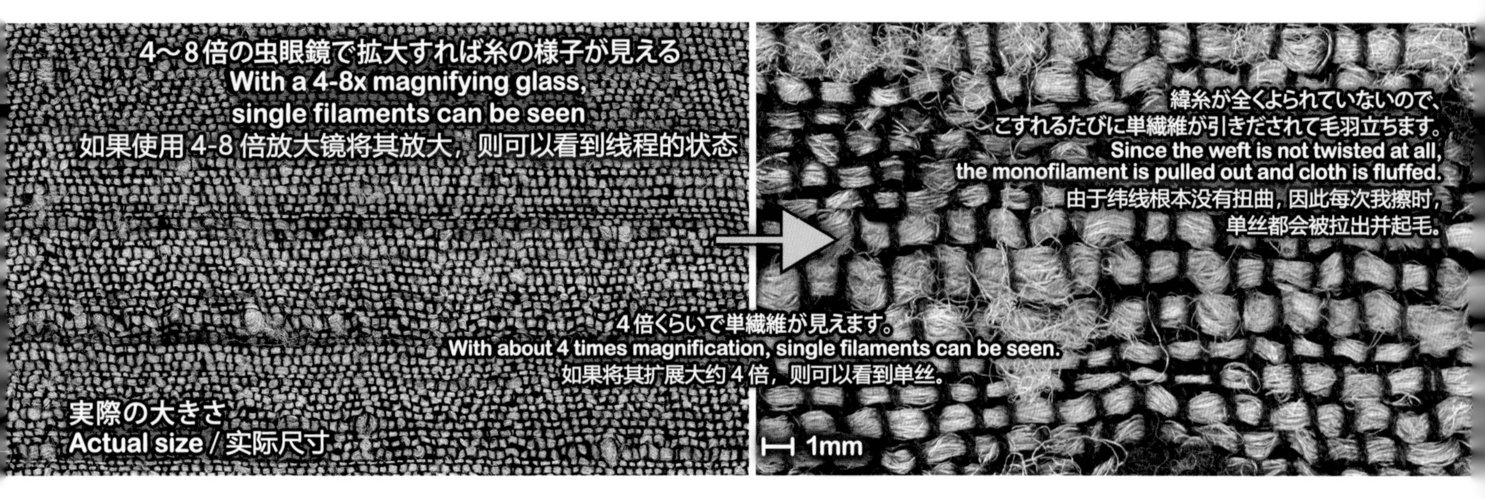

上の布はタッサーシルクです。緻密に織られていたので、まさか毛羽立つとは思いませんでした。見た目よりも薄くて軽く、気持ちのよいスラックスになると思ったら、着たその日に毛羽毛羽になりました。

インドではスラックスにしないのでしょう。ショールを作ればよかった。

糸が飛び出たところは糸を針で戻し、毛羽は少し刈り込んで、スチームアイロンをかけると目立たなくなりましたが、外出着にはなりません。

The upper cloth is tasar silk. Because it was woven precisely, I did not think that it would get fluffed. Since it looked thinner and lighter than it actually is, I thought a comfortable slacks could be made, however just on the day of first wearing, it got totally fluffed. I think this cloth are not used to make Slacks in India. I should have made a shawl.

At the spot where threads popped out, the thread was put back with needle and fluff was trimmed a little. After ironed with steam, fluff became inconspicuous, but the slacks couldn't be worn for outdoor cloth.

上面的面料是塔萨尔丝绸。因为它是精确编织的，我没想到为它不会变得起毛。可做看起来更薄，更轻且更舒适的休闲裤、但是那天我穿的时候它变成了起毛。我认为在印度，这面料是不会制休闲裤的。我认为做披肩比较好。

当线弹出时，用针将线放回去，将绒毛修剪一下，然后进行蒸汽熨烫时，它不起眼，但适合在家中穿。

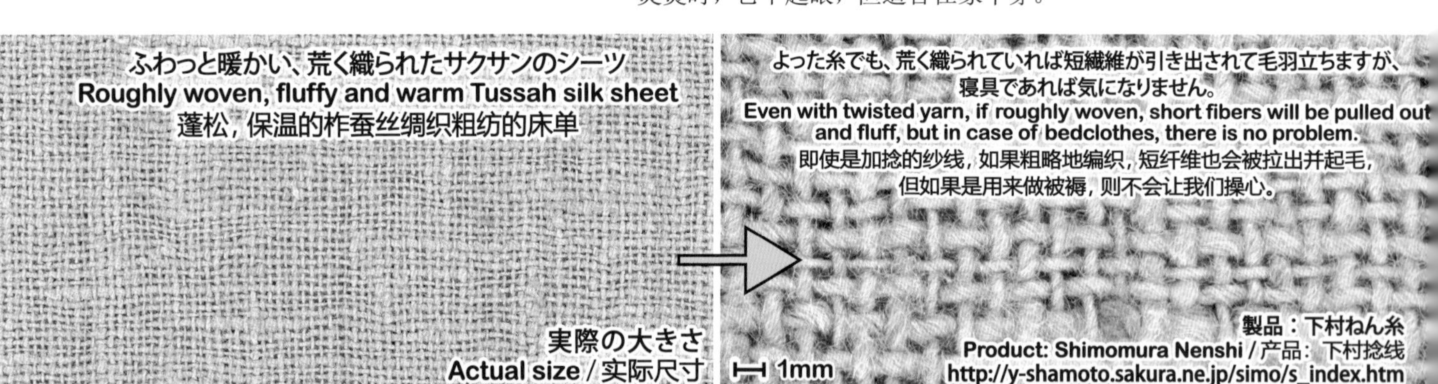

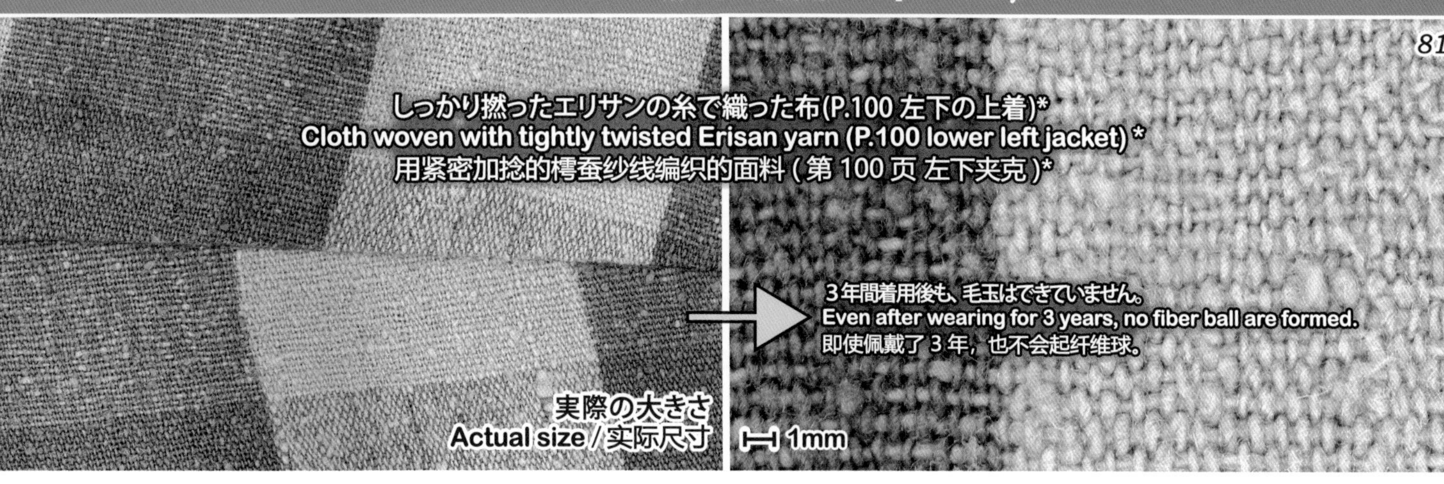

毛羽立ちやすい布 / Cloth easy to fluff / 容易起毛的面料

子ども時代に毎日のように使った、銘仙の座布団や縮緬の風呂敷は、毛羽立ちませんでした。しかし、きれいな刺繍や朱子の小座布団はこすれると毛羽立ったので、不思議に思いました。

2013年秋から絹の手編み糸を編むようになって、引っかかりやすい糸が多いことに驚きました。

The Meisen cushion and the FUROSHIKI of Chirimen (crepe), which I used every day in my childhood, did not get fluffed. However, the beautiful embroidery and the small cushion of satin fuzzy when rubbed, which made me wonder. Since autumn 2013, I started to knit silk hand knitted yarns, and I was surprised that many yarns are easily entangled.

我小时候每天都在使用的 Meisen 靠垫和绉绸的 FUROSHIKI 并没有起毛。然而，美丽的刺绣和缎子的小垫子在摩擦时起毛、这让我感到奇怪。从 2013 年秋天开始，我开始编织丝绸手工针编织纱线，令我惊讶的是许多丝纱线很容易被抓住。

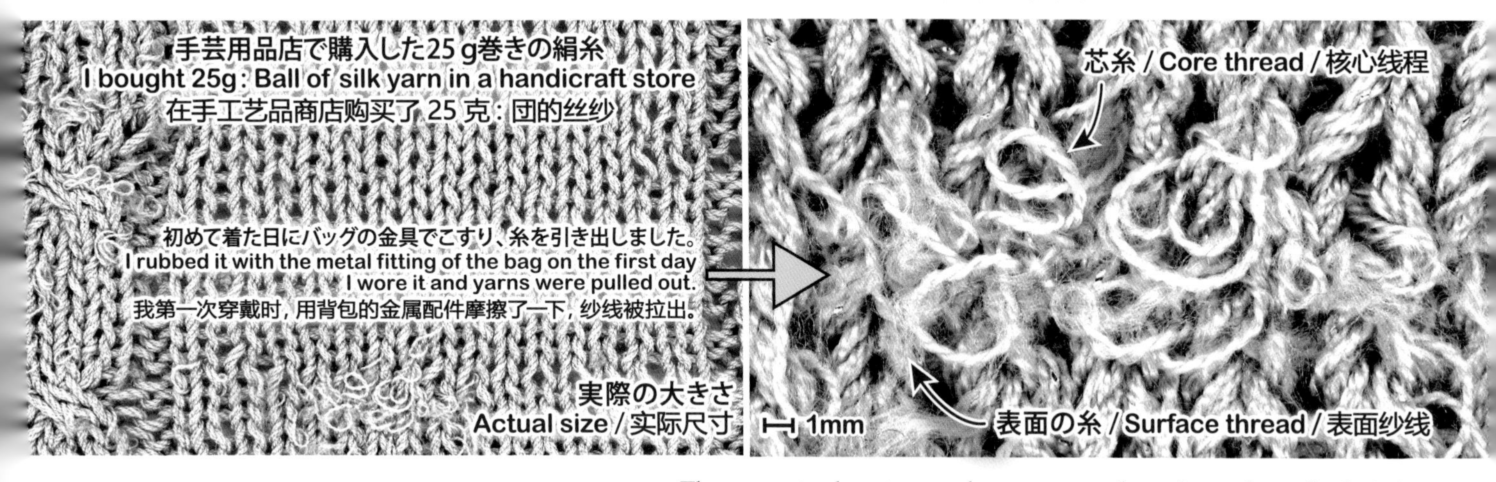

上の写真の糸は、ショールやテーブル敷きなどを編む糸のようでした。

毛糸や綿糸は予想通りの作品が編めます。しかし絹糸は、飾り物を編む糸と、日常着を編む糸が異なることを知りました。

絹のニット用の糸を販売している店舗はとても少なく、糸の種類も少ないので、私はニット用以外の織物用の糸も編んでみて、日常的に着用し、毛羽立ちや毛玉の様子を見ることにしました。

The yarns in the picture above seem to have been those for knitting shawls, table cloths, etc. Wool and cotton yarns can be knitted as expected. However, the silk yarn realized that the yarn knitting the decoration and the yarn knitting the sweater are different.

There are very few stores that sell silk yarn for knitting, and also there are few types of yarn, so I tried to knit yarns for textile other than for knit and to observe the occurrence of fluff and / or fiber ball while wearing daily.

上图中的线似乎是用于针编织披肩，桌布等的丝纱线。毛线和棉线可以随意编织。然而，丝线编织装饰物时和针编织日常衣服的纱线是不同的。

销售丝绸针织纱的商店很少，纱线的种类也很少，所以我决定为针织物以外的纺织品编织线，并每天穿着以观察是否起毛和起球。

＊エリサンの布（ショール）：ブータン製。現地購入品。

＊Eri silk's cloth (shawl): Made in Bhutan, locally purchased.

＊樗蚕丝的面料(披肩)：在不丹制造。在当地购入。

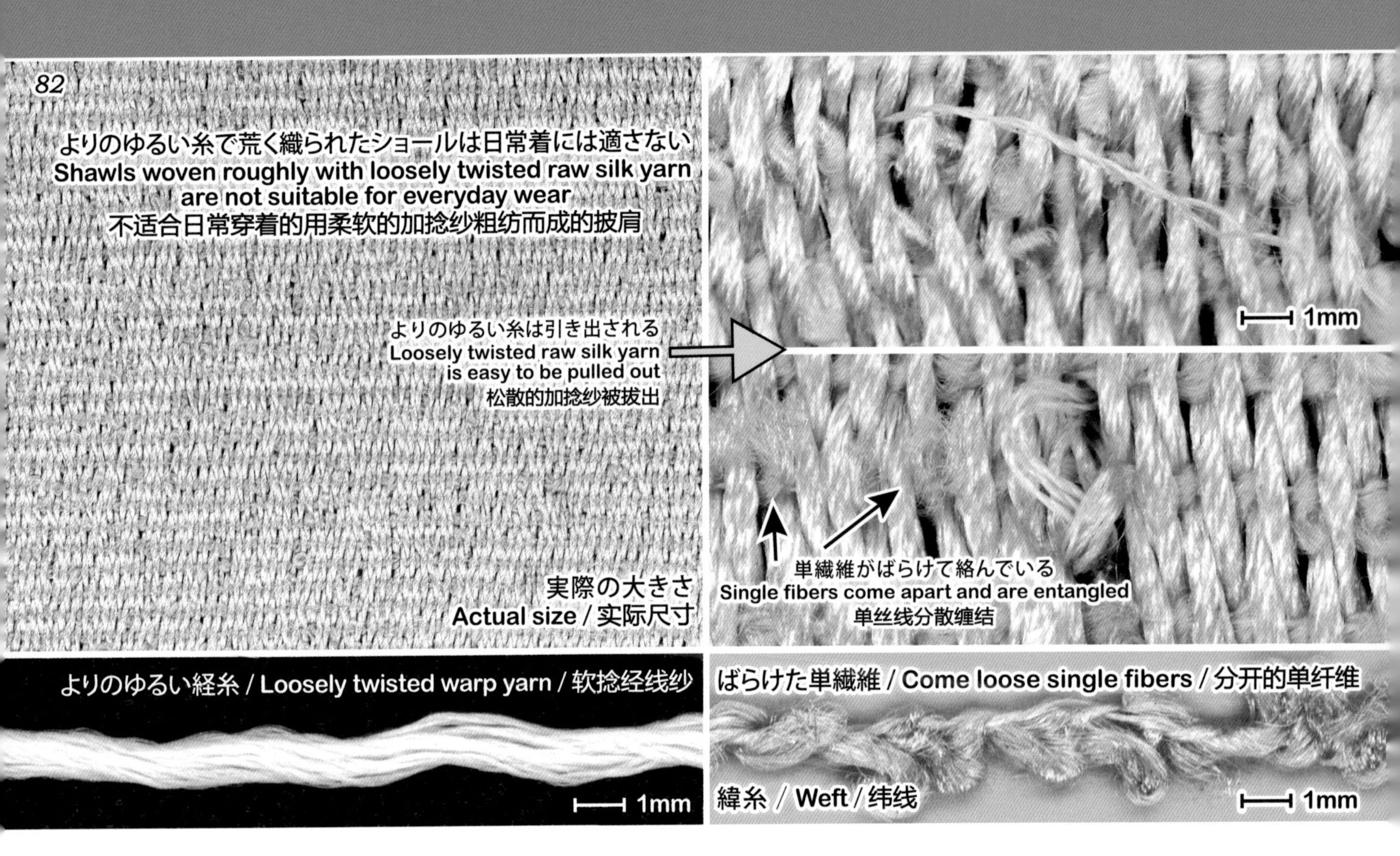

絹の布、ニット糸を探す / Find silk cloth, silk knit yarn / 寻找真丝布与丝针织纱

1980年代まで絹のスーツやコート地は販売されていました。絹の裏地は、2000年近くまで購入できましたが今では見かけません。

私は着物の羽裏やほどいた着物地を使っています*。

厚手の絹ニットも流通していません。私は60歳ころから毛糸が肌に刺さるように感じ、綿ニットを着ていました。絹に出会ってからは、ニット用の糸をホームページや手づくり市、帯や着物の手織工房などで探して編んでいます。絹の糸は生産量が少なく、次回は入手しにくいので、気に入った糸は多めに購入するように注意しています。

服地は自分で織るか、古い着物の生地や市販のストールなどを利用しています。

もう着ない着物は、パジャマやシーツにして直接肌に着ると、絹本来の心地よさがわかります。思い出の詰まった絹の布を活用した生活は、エコ以上に、気持ちのよい生活になりました。

Silk fabrics for suits and coats were generally sold until the 1980s. Silk lining could be purchased until nearly 2000, but it can't be seen now. I am using the back of kimono and the kimono fabric for silk lining*. Thick silk knit is also not circulated.

From around the age of 60, I felt the wool yarn pierce my skin, so I was wearing cotton knit. After I encountered silk, I have looked for it on website, handmade market, handwoven workshop, etc. and have knitted it. Since silk threads are low in production, it is difficult to obtain next time, so I am paying attention to buy a favorite thread a little more than the required amount.

Clothes are woven by myself, or I can use old kimono cloth or commercially available shawls.

For kimonos that I don't wear anymore, I can feel the natural comfort of silk by making pajamas and sheets and wearing them directly on my skin. Life with silk cloth full of memories is not only mere ecological but very pleasant.

西装和大衣的丝织物面料一直销售到1980年代。我可以在2000年左右，买到丝绸衬里，但现在看不到了。我正在使用和服的衬里面和和服面料＊。

厚实的丝绸针织也不流通。从60岁左右开始，我就感觉得羊毛纱线扎疼了我的皮肤，所以我穿着棉针织衫。遇到丝绸后我，我会在主页，手工市场，手工编织作坊等地方寻找，我針编织丝纱线。由于蚕丝线产量低，很难下次获得，所以我注意购买比所需数量多一点喜欢的线。

要么衣服面料是我自己编织的，或使用市场上的旧和服织物或披肩。

对于不再穿的和服，我可以制作睡衣和床单并将它们直接接触到我的皮肤上，从而感受到丝绸的原始舒适感。

布满记忆的丝绸生活既是一种宜人的生活，又是一种生态 。

*絹の裏地は、良質な裏地として使うほかに、裂き布として織物や編み物に使えます。

*Silk lining can be used not only as a high quality lining, but also as "cloth strips yarn" for textiles and knitting.

*真丝衬里既可以用作高质量的衬里，也可以用作纺织品和针织的【布条线】。

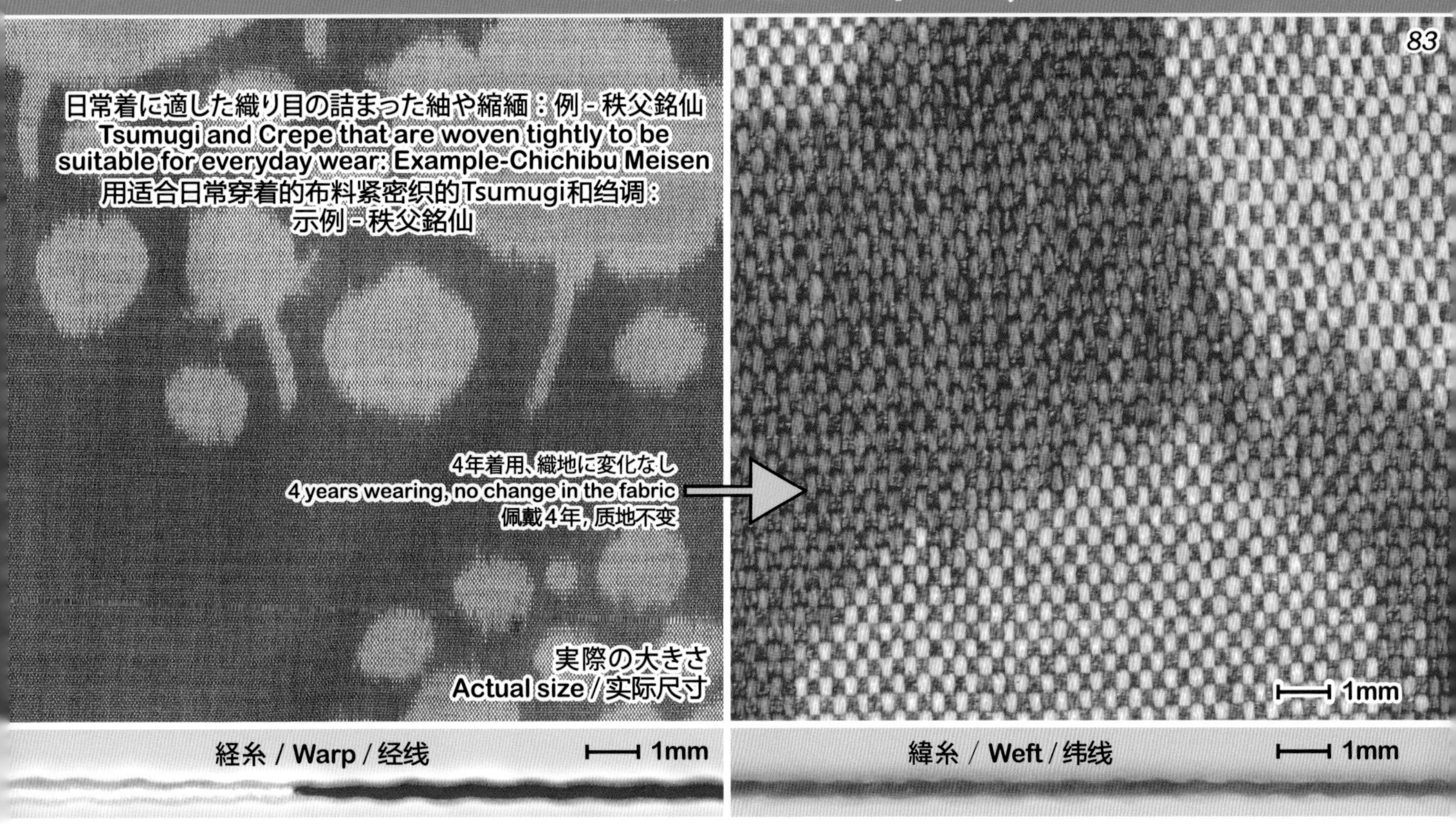

着物地をリユースするときの注意点
Points to remember when reusing kimono / 再用和服面料的注意事项

着物をリユースするときは、今後家庭で洗濯をして着るために、まず洗濯機で洗います*(反物の場合も)。特にちりめんやお召しのように、緯糸が強撚糸で織られた布は、水に入れると縮みます。洗ってからほどき、もう一度洗濯すると安心です。

着物をほどいたら、四角い布の集まりになります。ばらばらにならないように洗濯袋に入れて洗います。水を切るための脱水は、30 秒くらい。軽く水を切るだけ、それ以上脱水するとシワになります。半渇きになったら、スチームアイロンを表からかけて布を整えます。

絹の洗濯表示によるアイロンの温度は「低」ですが、低ではよれた繊維は戻りません。「高」にしてしっかりスチームを出してアイロンをかけます。小さな布で試してみましょう。スチームが出ていれば、アイロンがよくかかり、繊維は変質しません。美しい布が完成します。気に入った長さで切って、布はしを整えればショールになります。

When reusing kimono, it is first washed in the washing machine (including the roll of cloth)* in order to wash it at home in future. Especially the clothes woven with weft of hard twisted yarn such as crepe (Chirimen) and Omeshi, shrinks when they are placed in water. It's better to wash them again after once washed and unwound. When kimono is taken apart, it becomes a group of square cloths. Wash them in a washing mesh bag so as not to come apart. Dehydrate for about 30 seconds and drain. Just lightly cut off water, it will wrinkle if it dewater any further. When cloth is half dried, it is ironed with steam from the upper side and arrange the texture.

According to the laundry instruction for silk, the ironing temperature is "low", but with that, twisted fibers cannot be stretched. Set temperature to "high" and iron with enough steam. Let's try with a small cloth. Ironing can be done well without deteriorating fibers as long as steam is spewed out. A beautiful cloth is completed. If you cut it with the length you like and arrange the edges it will be shawls (P.26).

再用和服面料时，首先洗衣机中洗净，以便将来家中洗净和穿用(包括卷的布)*。尤其是用紧密的加捻纬线编织的面料，例如绉绸(Chirimen)和特等绉绸(Omeshi)，当放在水中时会收缩。将洗后和服解开接缝后，再洗即可放心。解开和服时，会得到集团一组方布。用洗衣袋洗涤，以免其散落。脱水约30秒并沥干。只是轻地切断水，如可过变地脱水，它就会起皱。半干后，用蒸气熨斗面上熨一下布即可平整。

真丝洗衣标注中熨斗的温度是“低”，但是在低时，扭曲的纤维不能被拉直。将其设置为“高”，并牢固地排出蒸汽并熨烫。用一块小布让我们尝试。如果蒸汽散发出来，它将被很好地熨平并且纤维不会变质。完成的美丽布。如果将其切成所需的长度并修剪边缘，即可成为披肩(第26页)。

*洗濯機で洗う：洗濯袋に入れて水を使用、温水の場合は、20～30度。中性洗剤を使用し、仕上げ剤や漂白剤は使いません。手洗いの場合は洗剤に浸けておいてから押し洗い。

* Wash in washing machine: Use washing mesh bag and wash with water or 20 - 30 degrees of hot water. Use neutral detergent, do not use neither finishing agent nor bleach. In case of hand wash, soak it in detergent and push it up and down.

*机洗: 使用洗衣袋。将水温调至 20 至 30 度的温水，使用中性清洁剂，请不要使用精製剂和漂白剂。如果要进行手洗，将其浸泡在添加了洗涤剂水軽中并軽柔推动洗。

手のかけ甲斐があるように / So that it is worth the effort / 喜欢的这样值得付出努力

手編み糸には毛羽立つ糸があります。

真綿を軽く撚った糸や撚られていないキビソ糸などは、柔らかい質感と暖かさを追求した糸です。こすれれば毛羽立ちます。以下は、とても毛羽立つ、サクサン手編み糸の経年着用の結果です。

Some hand-knitted yarns have fuzziness. Lightly twisted floss silk threads and untwisted Kibiso threads are ones that pursue soft and warm texture. It fluffs if rubbed. The following are the results of wearing fuzzy sakusan hand-knitted yarn over time.

一些手工编织的纱线有起毛。轻捻丝绵线和未捻 Kibiso 线是追求柔软质地和保暖性的线。如果揉搓，它会起毛。以下是长时间穿着模糊的柞蚕，手工针编织纱线的结果。

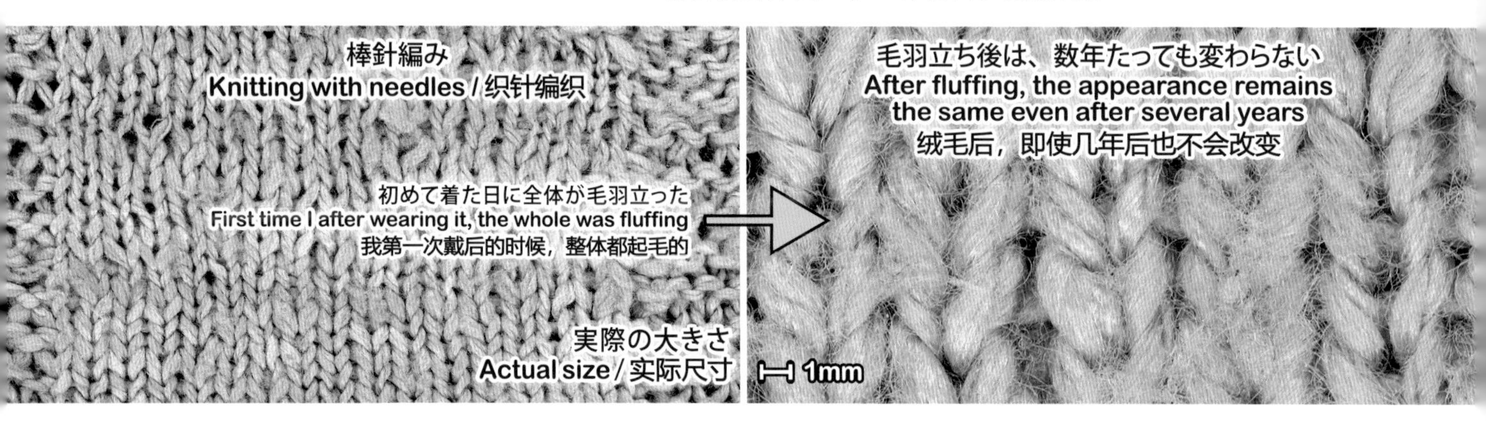

毛羽立っても暖かく、真冬でも夜着の上に羽織れば家事ができます。

病人が着替えず、家事が可能ならばすぐ休め、回復が早く自立できます。虚弱体質で、病気勝ちな私の有難い日常着です。

Even if it fuzzes, it is warm, and even in the middle of winter I can do housework simply by wearing a sweater over a night clothe. If sick persons can do homework without changing clothes, they can rest immediately and become independent soon owing to quick recovery. It's my grateful everyday wear because I have a weak constitution and often get sick.

它蓬松而温暖，即使在寒冬，我也可以在睡衣上穿一件毛衣来做家务。如果病人不换衣服可以做家务，他 / 她可以立即休息，可以加快康复并变得自立。我身体虚弱容易生病，因此我感恩于我的日常穿着。

毛羽立ちや毛玉を目立たせない編み方
Knitting technique for fluff and fiber balls not to be visible / 如何编织，使起毛与起球不很显眼

散歩着にもしたいので、試しにモチーフ編みにしたら、毛玉が目立ちませんでした。

穴あきだらけなのに冬は暖かく、夏は冷房よけになりました。

汗をかいた真夏の肌にも直接着られ、UV対策にもなるので、とても便利な一着です。

I wanted to use it as a walk clothes, so I made a motif knitting as a trial and the fiber ball did not stand out.

Although there are many holes, it is warm in winter and prevents excessively being chilled under air conditioning in summer.

It is a very convenient wear, because I can put it on directly on sweaty midsummer skin and it prevent UV ray.

我想将它用作步行服，所以当我尝试用图案花边编织时，起球就没有很突出。尽管洞很多，但仍冬天温暖，夏天凉爽。它非常方便，因为它可以直接戴在盛夏的出汗皮肤上并防止紫外线。

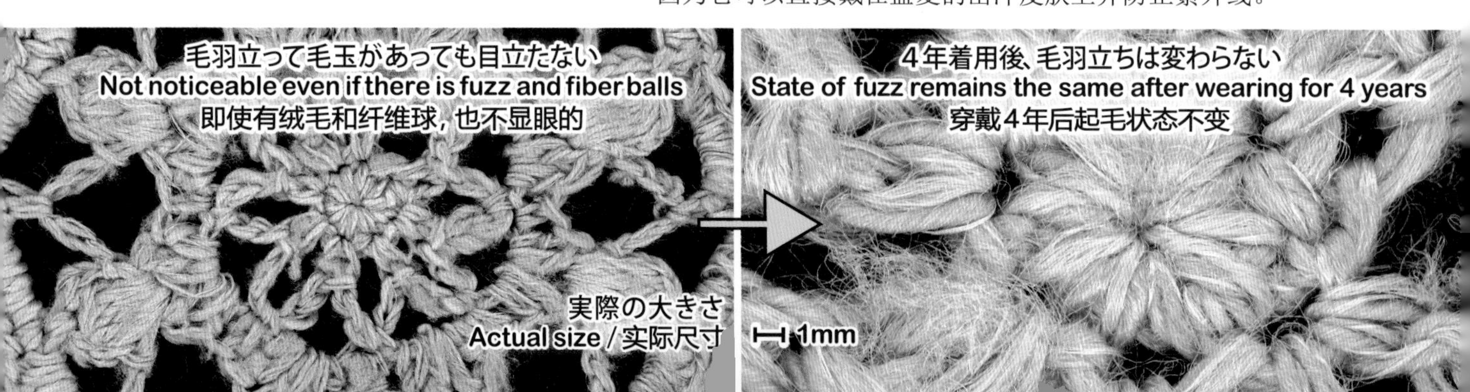

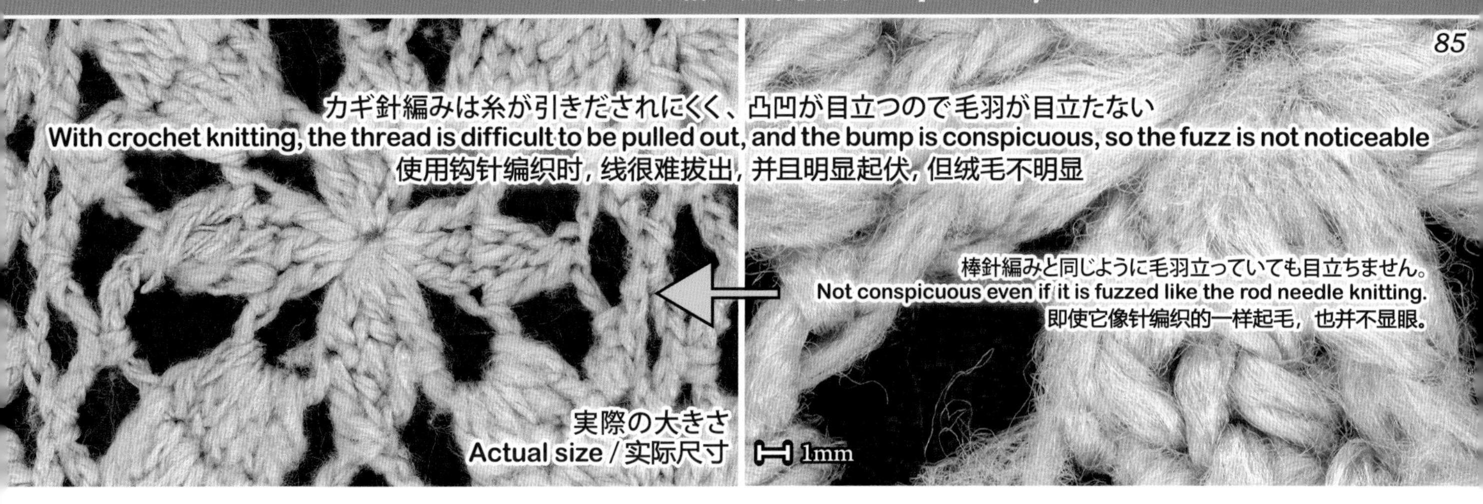

毛羽立たない糸を探す / Find a fluff-free thread / 寻材確无起毛起球的纱线

6年間毎月1kgくらいの糸を織ったり編んだりして着用試験をした結果、毛羽立つ糸と毛羽立たない糸は、作り方が違うことを理解しました。

絹の手編み糸の流通量は少なく、製作見本も少ないので、糸を購入するときには、セーターを編んだ場合の毛羽立ち方を販売者に確認することが重要です。

Weaving or knitting about 1 kg of yarn every month for 6 years, and as a result of the wearing test, we understood that fuzzy and non-fuzzy yarns are made differently. Circulation of silk hand-knitted yarn is small and there are few production samples, so it is important to confirm to the seller how fuzzing occurs at the time of knitting a sweater, when yarn is purchased.

连续6年每月织编与手工编织约1公斤纱线，和身上穿着面料测验结果，让我们了解到蓬松纱线和非蓬松纱线的制造方法不同。丝绸手工编织纱线的流通量很小，生产样品很少，因此，必须与卖方确认购买纱线时针织毛衣时，是如何蓬松纱线的起毛。

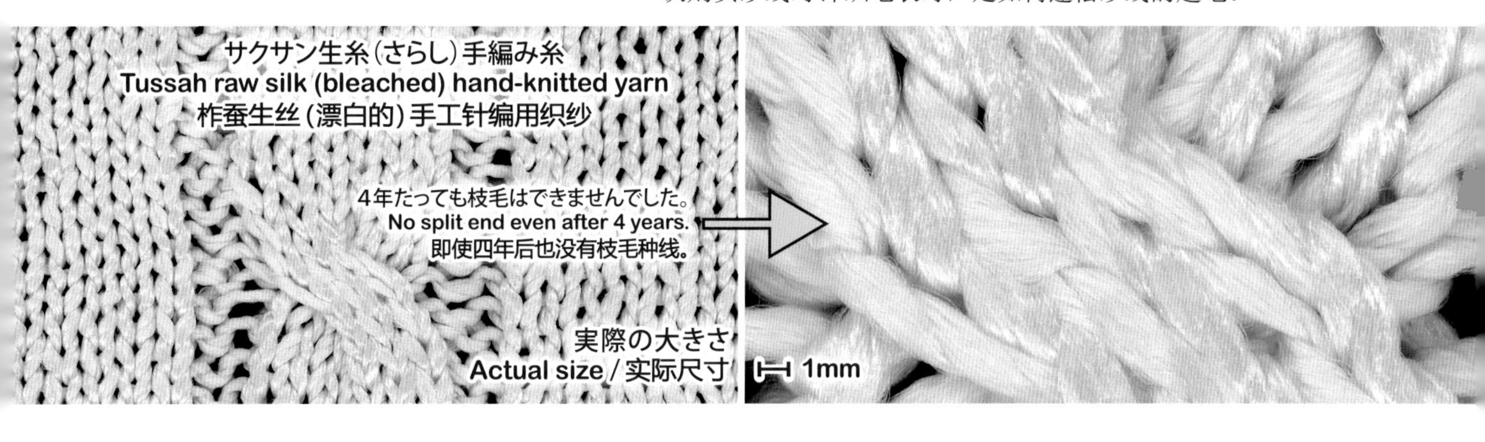

上の糸は 2015 年、山形県山辺町の絹織物の専門店「まゆや あだち」*を訪問したときにお預かりしたサクサン糸です。600g弱で、バルキーセーターを編みました。

太糸なので、81ページ右下のニットのようになると思いましたが、毛羽立たず何年たっても新品のように着ることができます。

このような糸もあるのだと思い、更に色々な糸を編みました。

Upper thread is the Tussah raw silk that I received when I visited Mayuya Adachi*, a silk fabric store in Yamanobe Town, Yamagata Prefecture, in 2015. With a little less than 600 g, I knitted a bulky sweater. I thought it would be fuzzing like the knit on the lower right of page 81 because it is thick thread, but you can wear it like a new one even after many years without fuzz.

I thought that there is such a yarn, and have knitted various threads further.

上面的线是2015年我拜访Mayuya Adachi*（我在山形县山之边市的一家丝织物专卖店）时我收到的是柞蚕生丝。用不到600克的重量，我编织了一件宽松的毛衣。

我以为它会像第 81 页右下角的针织物一样起毛突出，因为它的线很粗，但即使经过多年却不贝起毛，仍可以像新的一样穿上它。

我以为有这样的线，进一步针编织各种线。

*安達㈱：http://www.a-mayuya.jp

*Adachi Co.: http: //www.a-mayuya.jp

*安達株式会社:
http://www.a-mayuya.jp

太いサクサン生糸をかぎ針で編んだポシェット
Thick Tussah raw silk yarn pochette knitted with crochet needle
钩针编织的厚实柞蚕作生丝纱线的小挎包

1年間毎日使用しても、毛玉はありません。
Even I use it every day for one year, there is no fiber ball.
即使我每天使用一年都没有纤维球。

キビソ糸で編んだネックウオーマー
Neck warmer knitted with Kibiso thread
带Kibiso线编织的暖颈套

経年使用で知ったこと / What I learned from many years use / 我从多年使用中学到的东西

前ページ下のセーターとおそろいで上のポシェットを編みました。1年中使っても毛玉ができず、こすれに強いことに驚きました。

不思議に思い、安達㈱*に問い合わせたところ、35中のサクサン糸を132本、4回逆向きに撚り合わせた糸なので、単繊維が飛び出しにくいと、安達社長から説明を受けました。

滋賀県長浜市の㈲きぬさや**に、工房風花で購入した糸について、質問をしたときにも、しっかり撚ってストレスをかけた糸は、枝毛と毛玉ができにくいと説明されました。

そのため、縮緬のように数千回撚った糸で織ると毛羽立ちにくいのでしょう。

また真綿を撚らず、同じ太さに引き出しただけの糸を、強く打ち込んで織る結城紬は、糸がしまって、毛羽立たないのだと考えられます。

I knitted a pochette (photo above) matching the sweater of the previous page. I was surprised to know that it was strong against rubbing, because fiber ball did not appear during my one year usage.
With wonder, I inquired to Adachi Co., Ltd.* and President Adachi answered that the single fiber is difficult to jump out because 132 yarns of 35 M Tussah silk yarn were twisted in reverse direction four times.
When I asked Kinusaya Inc.** in Nagahama City, Shiga Prefecture about the yarn purchased at Atelier Kazahana, it was explained that the yarns that were twisted and stressed firmly are apt not to generate split ends and fiber balls.

Therefore, when woven with yarns twisted several thousand times like Crepe (Chirimen), it would be hard to fuzz. Also, in case of Yuki Tsumugi which is pulled out while maintaining the same thickness without twisting the floss silk, fuzzing would be difficult to happen if the yarn is tightened by the strong weft-shooting operation at the time of weaving.

用与上一页的毛衣的相同纱线，我编织了上图中的小包包。我整整用了一年，甚至都没有出现纤维球。它很强的抗摩擦性使我感到惊讶。我想知道，当我联系Adachi公司*时，Adachi社长回答说，单根纤维没有跳出来是因为132根35M纱线的柞蚕丝被反向扭转了四次。当我向滋贺县长滨市的【Kinusaya**】询问在工房風花购买的纱线时，有人解释说的加捻和受力纱线过大的纱线很难起毛和起球。因此，如果您用像绉调纱(Chirimen)那样扭曲数千次的纱线织编，就很难起毛。此外可以考虑，与Yuki Tsumugi一样，即使即使在不扭曲真綿丝线的情况下将纱线拉至相同的厚度纱线，如果在编织时收紧纱线，单根纤维也很难跳出。

かぎ針編みなので、糸が引き出されにくく、毛羽立ちしにくい
Since it is crochet, it is difficult for the thread to be pulled out and fuzzing is difficult to occur
由于是钩针编织，因此很难拔出线，并且不易产生绒毛

1mm

こすれないので5年間着用後、あまりかわらない
It won't rub, so it won't change
changes sharply, and so the finishing appearance of knitted products looks very interesting.
没有摩擦，因此穿戴5年，也不会有太大变化

キビソ糸***は、糸の太さが極端に変化するので、編み上がりが面白い。
In case of Kibiso yarn***, the thickness of the yarn changes sharply, and so the finishing appearance of knitted products looks very interesting.
Kibiso纱线***具有粗犷的针织效果，因为纱线的粗细会急剧变化。

1mm

*安達㈱：http://www.a-mayuya.jp
**㈲きぬさや：http://www.brandnewsilk.com
***糸：工房 風花 P.117

*Adachi Co. Ltd.: http://www.a-mayuya.jp
**Kinusaya Inc.: http://www.brandnewsilk.com
***Thread: Atelier Kazahana P.117

*安達株式会社：http://www.a-mayuya.jp
**Kinusaya Inc.: http://www.brandnewsilk.com
***线：工房 Kazahan P.117

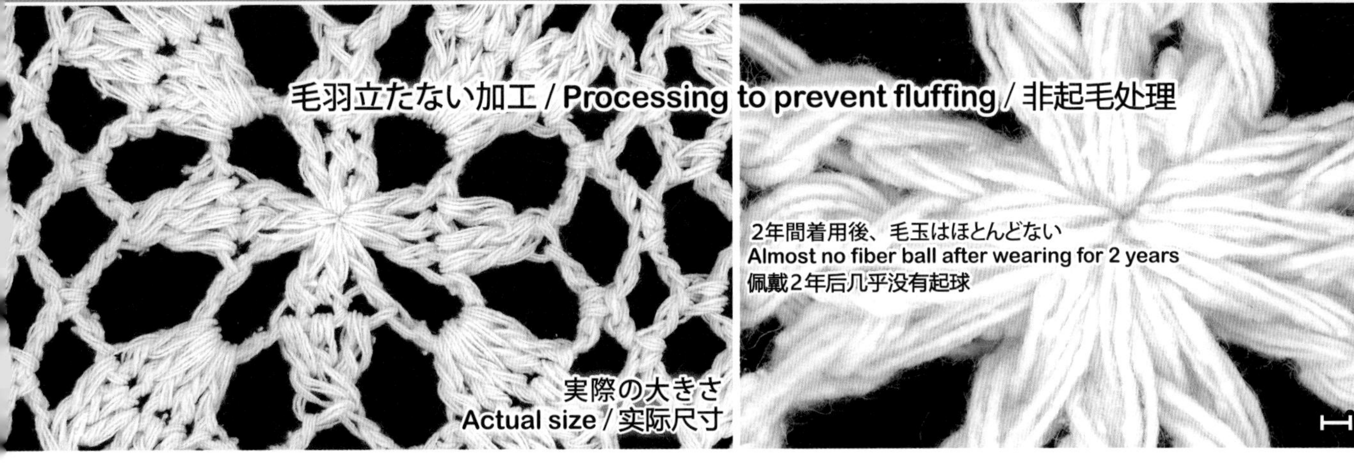

ある日、札幌の手織工房で絹糸の擬麻(ぎま)加工に出会いました。その不思議な質感の糸を、都内でも販売する店舗*があることを知り、早速手に入れました(上掲)。擬麻加工とは、セルロース系の樹脂をコーティングして麻に似せ、毛羽立ちを押さえて糸にハリを出す加工です。編み進むとコーティングしたセルロース樹脂が少しはがれ、粉落ちするのが気になりました。サマーセーターなどに、さらっとした気心地が好まれるとのこと。毛羽立つサクサンのモチーフと交互に編み込み、1年中着ていますが、数年後も毛羽立ちが気になりません。

最近、バルキーセーターを編んで毛羽立ちが少なかった糸は、工房風花の絹紡紬糸や生糸を引き揃えた糸。下の写真のように、しっかり撚った絹紡紬糸の細糸と真綿糸の細糸が混ざると、彩りが面白い。もちろん右下の写真のような、太さ2mmくらいのサクサン生糸**や、絹紡糸の2本どりも、安心して使えます。

One day, I met a imitation linen finish of silk yarn at Sapporo's hand - woven atelier. I learned that there is a store * that sells that curious texture of thread in Tokyo, I got it immediately (above). Imitation linen finish is imitating hemp by coating cellulosic resin, to restrain fuzz and give tension to the thread. As I knitted, I was worried about the coated cellulose resin peeling off a little and the powder falling off. As a summer sweater material, its dry feeling is said to be preferred. It is knitted alternately with the fuzzy tussah silk motif, and I wear it all the year around, but I don't mind fuzzing even after several years.

Recently, the yarn that had less fuzzing when knitting bulky sweaters are knitted, is a yarn made up of aligned silk-spun yarn, etc. made by Atelier Kazahana **. As shown in the picture left below, the colour combination is interesting when the fine thread spun silk threads and the pure floss silk threads are mixed.

Of course, you can also safely use 2-strand of the 2 mm thick tussah raw silk** or silk spun yarn as shown in the lower right picture.

有一天，我在札幌的手工编织车间里遇到了一种仿亚麻加工的丝绸线。我了解到东京有一家商店*销售这种神秘的纱线，我立刻去买到了。仿亚亚加工是涂覆纤维素树脂以模仿麻并抑制起毛以赋予线弹性的过程。针编织时，我担心的是涂覆的纤维素树脂脱落一点并散落。作为夏季毛衣的材料，据说由于其干燥爽感而优选加工。我编织图案花边针织物，和起毛的柞蚕丝绸交替编织，我一年四季都穿，即使穿过了几年也我不会担心起毛。

近来，在编织大件毛衣时，起毛较少的纱线是由工房Kazahana生产的对齐的丝绵线和绢紬丝线纱组成的纱。如左下图所示，将细线真丝纺和绢丝纱混纺时，颜色组合很有趣。当然，您也可以安全地使用2股2毫米厚的柞蚕生丝**丝或丝纺纱线，如右下图所示。

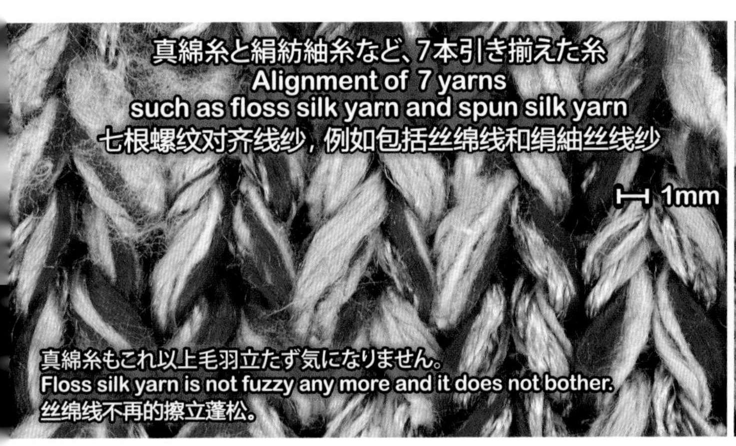

*糸：東京アートセンター：https://www.artcenter.co.jp
**糸：安達㈱

*Tokyo Art Center: https://www.artcenter.co.jp
**Thread: Adachi Co. Ltd.

*Tokyo Art Center: https://www.artcenter.co.jp
**线：安達株式会社

シルクを編むときの注意点、ウールとの違い
Notes on knitting silk, difference from wool / 针编丝绸时的注意事项，与羊毛的区别

シルクは、ゴム編みの立体感がなく伸縮性がありません。
このレッグウオーマーは、ゴム編みに伸縮力がないので、足首周りや、
ズボンの上から重ねばきしています。
Silk does not have spatial effect of rib stich and is not elastic.
Since silk leg warmer has no elasticity when knitted in rib stitch,
it is worn around the ankle or over the pants together.
丝绸没有橡胶编织的三维效果并但且没有弹性。由于这种腿套在橡胶编织
中没有弹性，因此将其穿戴在适用于脚踝周围或裤子上方。

並太毛糸のレッグウオーマー、1組130g
DK wool yarn leg warmers, 1 pair 130g
粗毛线的暖腿套，一对130g

毛糸は10年後でも
ゴム編みの立体感は変わりません。
Three-dimensional feeling of
wool knitted in rib stitch does not
change even 10 years later.
羊毛纱线的橡胶编立体感，
即使10年也不会改变。

合細毛糸位の太さのサクサン糸*の2本どり、1組78g
Lace thick 3PLY of Tussah silk * 2-strands, 1 pair 78g
细线3PLY的柞蚕丝纱*2股，一对78g

5歳で編み物をはじめた私は、セーターをほどく手伝いをして、着るものの形を知りました。成人後イギリスやイタリアの糸が気に入って編みましたが、絹100%の糸には出会いませんでした。

絹は木綿よりもゲージを押さえにくく、着丈が10cm、袖丈も7～8cm伸びることがあります。

変形を知るため、試し編みをスチームアイロンで伸ばしてゲージを記録します。もしも、ふっくらとした編み目のままゲージをとれば、完成した日に、編み目のゆとりの部分がだらっと伸びます。毛糸を編むよりも細い針で、目を詰めて編むと、ほぼゲージ通りに編めます。しかし、人が着ると体温と湿り気で、5%くらい伸びるので、伸びることを考慮して編むと安心です。

I started knitting at the age of five, and while helping my mother untie a sweater, I learnt the shape of what I wore. After I grew up, I liked the British and Italian threads and knitted them, but I didn't encounter 100% silk thread.

Silk is more difficult to hold down the gauge than cotton, and it could happen that dress length extended by 10 cm and sleeve length, by7-8 cm. To know the deformation, the trial knit is stretched with a steam iron and the gauge is recorded. If you take a gauge with a plump stitch, the loose part of the stitch will stretch loosely on the day of completion. If silk yarn is knit tightly with a needle finer than the one used to knit the wool yarn, it can be knit almost exactly to meet the planned gauge. However, when a person wears it, it grows about 5% due to body temperature and dampness, so we should knit it in consideration of the extension.

当我五岁开始编织衣服时，我帮妈妈解开了一件毛衣，所以我知道穿的衣服的形状。我长大以后，我喜欢用英国和意大利的线并进行手工编织，但我从没遇到过100%的丝线。丝绸比棉花更难以掌握尺寸，其上衣长度可能会伸缩10厘米袖子长度可能会伸缩7–8厘米。要了解变形，用蒸气熨斗拉伸试编织并记录其标准针数。如果我规范使用的丰满针，则针头的松散部分会在编织好的当天会出現松弛延长的現象。如果使用的针比用于编织羊毛纱线的针细并紧密编织，则编织好的衣服尺寸几乎完全和针规一样。但是，由于的丝绸针织物在人体温度和湿度的影响下拉伸约5%，因此要考虑到其伸缩性的情况下进行编织。

ゴム編みにストレッチヤーンを
引き揃えて編みました。
Stretch yarn was knitted with
silk yarns together by rib-stich.
我用对齐的弹力丝编织罗纹针织。

P. 85 下のセーターと同様に4号針で編みました。
Like the sweater of P85 below, I knitted it with the No. 4 needle.
我像按第85頁下的毛衣一样用4号针织的。

*糸：下村ねん糸 P.80
染色：ワークショップにて、㈱みなみ紬
（鹿児島県奄美市）

*Thread: Shimomura Nenshi P.80
Dyeing: At the workshop of Minami Tsumugi Co., Ltd. (Amami, Kagoshima).

*线：下村捻线 第80页
染色：在Minami Tsumugi株式会社的车间，(鹿儿岛县奄美市)。

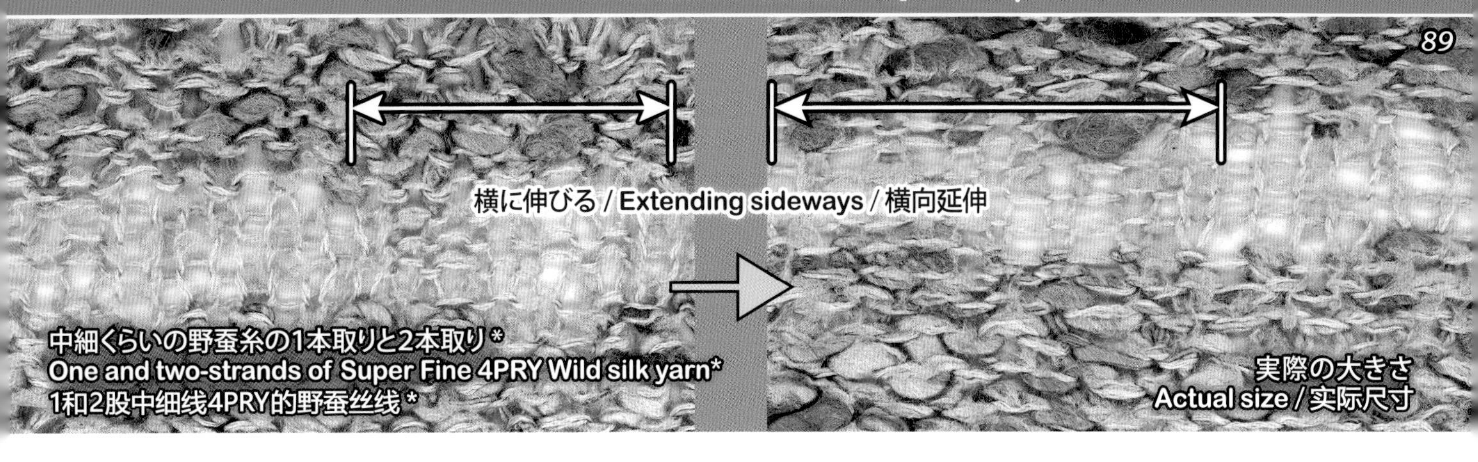

編み方で横に伸びたり、たてに伸びたり
Depending on the knitting method, it stretches horizontally or vertically / 根据编针织方法的,它水平或垂直拉伸

絹を編み始めたころ、寸法の押さえ方をメーカーに訊ねたところ、絹は寸法が押さえにくいので、バルキーニットの既製品は作らないと聞き、自分で実験しようと思いました。

綿糸を添えて編むと風合いを保ちやすいように思って編んでみましたが、ゴム編みが益々横に広がり、糸が重くなって全体に伸びました。編み方によって伸び方も変わります。

上のように表編と裏編みを数段ごとにくりかえせば横に伸びやすく、下のようなゴム編みはたてに伸びやすくなります。左ページの左下のように、ゴム編みにストレッチヤーンを引き揃えて編むと伸縮しますが、アイロンで溶けるので、ゴム編みの付近はアイロンがけに注意が必要です。

At that time when I started knitting silk, I asked the manufacturer how to fix dimension and was explained that Silk is difficult to fix the dimensions, and so bulky knit ready-made items are not manufactured. Then I decided to make an experiment by myself. I tried aligned cotton yarn to make it easy to keep texture, but the rubber knitting spread loosely sideways, and the thread became heavier and stretched through. How to extend also changes depending on the knitting method. In case knit and purl are repeated every several steps as in the picture above, it will be apt to stretch horizontally, and in case of rib-stich as in the picture below, it will be apt to stretch vertically.

As is shown in the lower left of the left page, rib stich of silk yarn aligned with stretch yarn gives elasticity, however care must be paid not to iron the part of rib stich because ironed part may melt.

当我开始针编织丝绸时，我问制造商如何匹配尺寸时，他们说“编织丝织物时很难掌握尺寸，因此他们不生产粗重的针织现成产品”，因此决定尝试。将我用对齐棉线编织以保持质地，但是橡胶针织渐渐地向横向散开，過线变得较重并遍及整体且一直延伸。根据编针织方法伸缩性也是变化的。如果如上图所示每隔几步重复进行正针织和反针织，则水平拉伸将更容易，而下图中的橡胶针织将更容易垂直拉伸。

正如在左页左下方所看到的那样，弹力纱与橡胶针织物对齐以进行膨胀和收缩，但是由于它会蒸气熨斗融化，因此注意不要在橡胶针织附近熨烫。

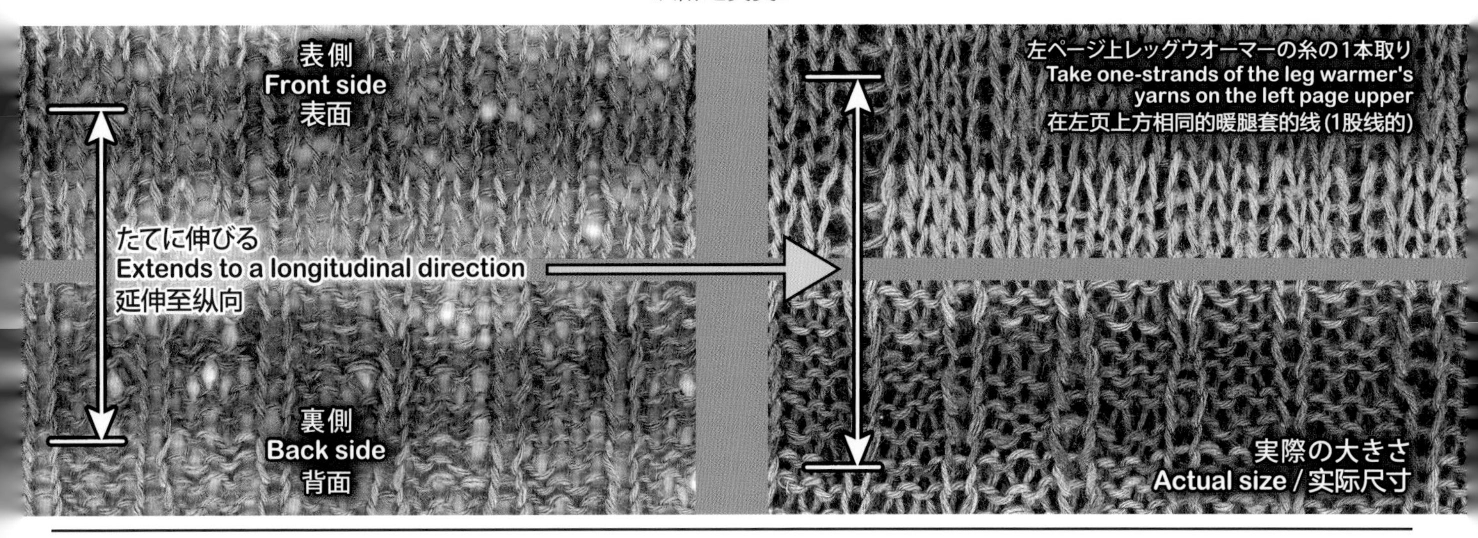

*野蚕糸：アトリエ トレビ (P.134)　*Wild silk thread: Atelier Trevi (P.134)　*野蚕丝线：Atelier 特雷维 (P.134)

生糸を編む / Knitting raw silk / 针织的生丝

太繊度低張力繰糸機で作られた1000デニールの生糸
1000 denier raw silk made with a high fineness, low tension reeling machine
用高纤度，低张力缫丝机制成的1000旦尼尔生丝

一度に300粒以上の繭を低張力で繰糸して作った1000デニールの太糸の美しさに見とれて編みました。
1000デニールの、未精練のキビソ糸は小枝のように固く、編みにくいのですが、バッグや帽子を編むことは可能です。
しかし生糸は、鋼のように編み針をはじき返します。10cm 四方の見本を編むのに四苦八苦しました。
そこで、熱湯に15秒くらい浸したところ編めるようになりました（右ページ画像参照）。

**Attracted by the beauty of 1000 denier thick raw yarn made by reeling more than 300 cocoons with low tension at a time, I knitted.
1000 denier un degummed kibiso yarn is hard like a twig and difficult to knit, but it is possible to knit bags and hats.
But raw silk repels the knitting needles like steel. I had a hard time knitting a 10 cm square sample.
So, when I soaked it in boiling water for about 15 seconds, I could knit it (see the image on the right page).**

一次以低张力一次缫丝300多茧制成的1000旦生丝的漂亮程度吸引了我也開始用此来。1000旦未脱胶的kibiso纱线像树枝一样坚硬，难以编织，但可以编织的袋子包和帽子。但是生丝会像钢一，和样排斥编织针。我也难编织一个10平方厘米的样品。
因此，当我将它浸泡在沸水中约15秒时，就可以手工编织了(请参见右图)。

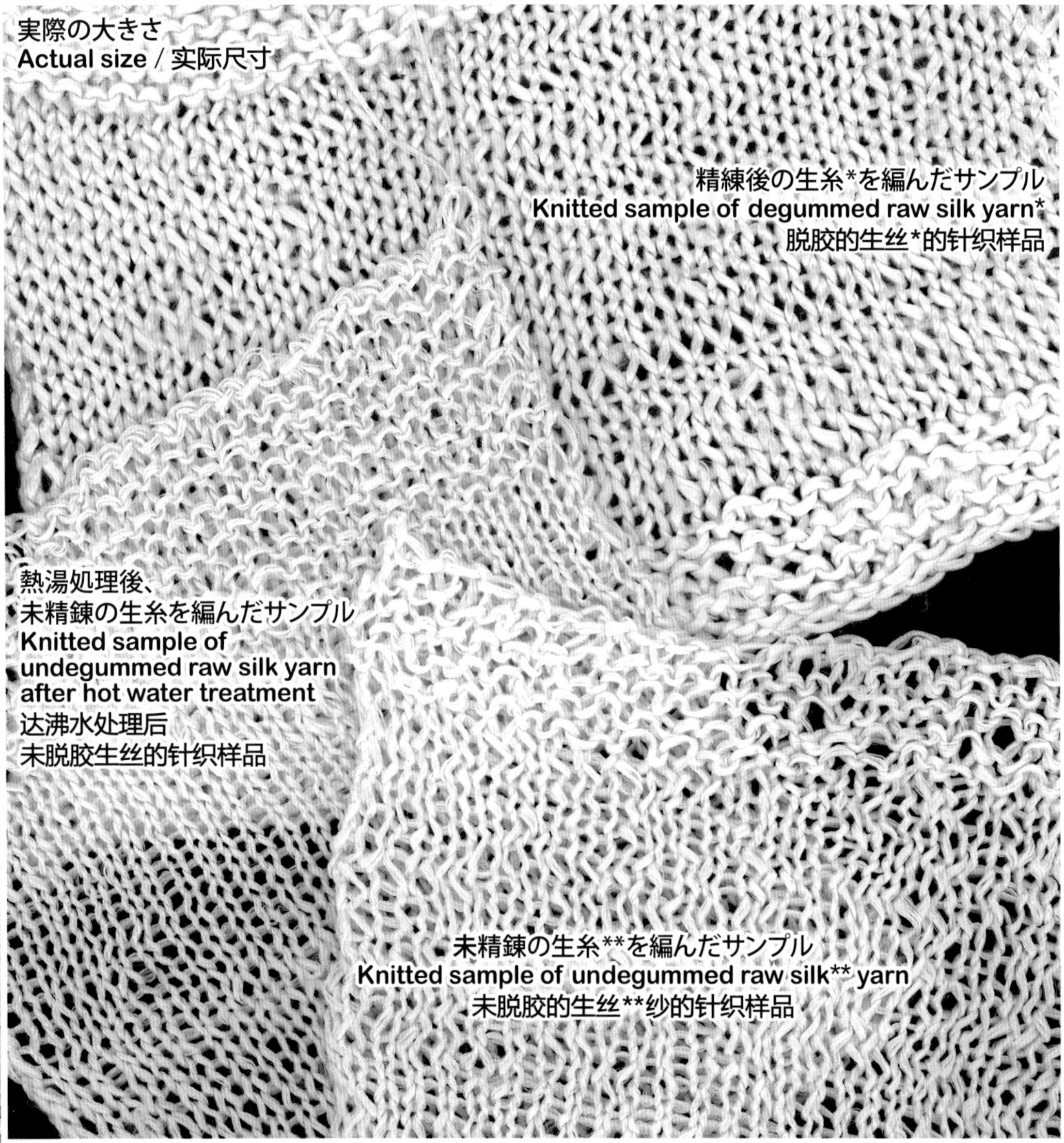

实際の大きさ
Actual size / 实际尺寸

精練後の生糸*を編んだサンプル
Knitted sample of degummed raw silk yarn*
脱胶的生丝*的针织样品

熱湯処理後、
未精錬の生糸を編んだサンプル
Knitted sample of undegummed raw silk yarn after hot water treatment
达沸水处理后
未脱胶生丝的针织样品

未精錬の生糸**を編んだサンプル
Knitted sample of undegummed raw silk yarn**
未脱胶的生丝**纱的针织样品

*糸：工房 風花
**糸：㈱宮坂製糸所（長野県岡谷市「岡谷蚕糸博物館」内）http://miyasakasilk.com

*Thread: Atelier Kazahana
**Thread: Miyasaka Silk Mill Co., Ltd. (in Okaya Silkworm Museum, Okaya, Nagano) http://miyasakasilk.com

*线程：工房 Kazahana
**线程：宫阪丝绸厂株式会社（在长野县冈谷市冈谷蚕糸博物馆）http://miyasakasilk.com

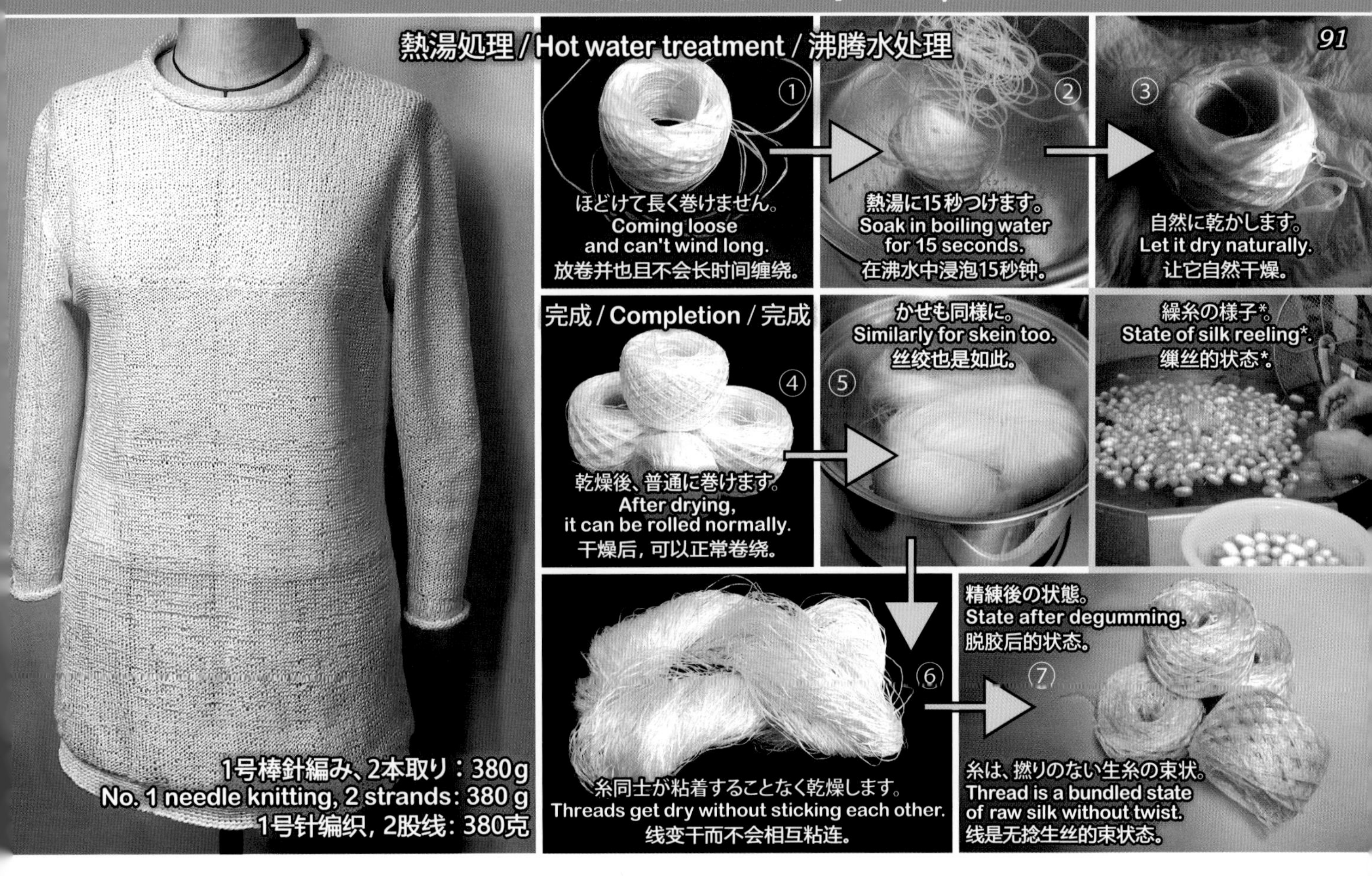

未精練生糸を編む方法 / How to knit undegummed raw yarn / 未脱胶生丝的手工编方法

生糸は通常精練後に織られますが、オーガンジーは生糸で織ります。セリシンにおおわれたままの硬い糸で織られた布は、透け感があって張りのある薄い布になり、フォーマルドレスや舞台衣装など、様々な美しいドレスを生みだします。

また、羽二重も生糸で織りますが、布にしてからセリシンを落とし、セリシンの厚み分が布のゆるみとなって、ふんわりとした布になります。

未精錬の生糸を編むことは不可能だと思いましたが、熱湯に浸すことで編めるようになりました。上の写真のセーターの透けた感じのところは、ぬれたままの糸を編んだ場所です。

300本以上の生糸が撚られずに並んでいるだけの糸なので、着用したらどのようなことになるかは、今後実験していきます。

Raw silk is usually woven after being degummed, organdy is woven in raw silk. Cloth woven with the hard thread that is still covered with sericin becomes a thin cloth with a feeling of translucence and tension, and produces various beautiful dresses such as formal dresses and stage costumes.

In addition, we also weave Habutae with raw silk, but after making it into cloth, remove sericin, and lost sericin layer make some space, so that cloth gets fluffy and soft.

I thought I couldn't knit undegummed raw silk, but I could knit it by soaking it in boiling water. The part of sweater which looks transparent on the picture above is the place where wet yarn was knit, and it is in a dry state. Since more than 300 raw threads are lined up without being twisted, I will carry out experiments to see what happens when I will continue to wear.

通常将生丝脱胶后编织，但是，将蝉翼纱布(organdy)是用生丝编织的。用含丝胶的硬纱线编织的面料有透明感和張力感的薄布，可用来生产各种漂亮的礼服，例如礼服和舞台服装。另外，电力纺(Habutae)生还用生丝编织而成，但是将其制成面料后，脱胶加工后，丝胶的厚度会使面料变松，从而形成柔软的面料。

我以为未脱胶的生丝，无法用于手工编织但可以将其浸泡在沸水后进行针织。将在上图中的毛衣的感觉透明感部分，是编织湿纱的地方和，还有它处于干燥状态。由于300多个生丝螺纹排成一列而没有被扭曲，因此我们将继续进行实验以观察继续佩戴时会发生什么。

---

*㈱宮坂製糸所：銀河シルク繰糸と呼ばれます。前頁の工房風花の糸と、上の写真⑦の糸は同じ機械を持つ、群馬県の㈱碓井製糸製です。

*Miyasaka Silk Mill: It is called galaxy silk reeling. Atelier Kazahana thread on the previous page and the thread in photo ⑦ above are made by Usui Silk Mill Co., Ltd. in Gunma Prefecture, which has the same machine.

*宫坂丝绸厂株式会社：它被称为银河丝缫丝。前一页的工房Kazahana线和上图⑦中的线是由群马县的Usui Silk Mill Co.,Ltd.制造的，它具有相同的机器。

仕事仕舞いの糸 / Yarn at closing the atelier / 关闭工作室后剩余纱线

色とりどりで長さもいろいろ* / Multicolored and various lengths* / 多种色彩色与各种长度*

WILD SILK MUSEUM 監修 / **WILD SILK MUSEUM supervision** / 野生丝绸博物馆监督

友人から贈られた絹糸
Silk yarn given by a friend
朋友赠送给的丝纱线

友人の大学の友人が、織らなくなったから、著者の中山に有効使用を望み託されたとのこと。
A friend of his friend's university sent it to the author, Nakayama, because his friend wove no longer and wanted him to use it for something effectively.
他朋友大学的一个朋友不能编织的丝纱线寄给了作者中山，希望他能有效地利用它。

＊太さも数種類あるので、太さごとに色のバランスをとりながらつなげ、7〜10本まとめてから撚りをかけました。

*Since there are several types of thickness, we decided to connect them while balancing the colors for each thickness, and twist 7 to 10 pieces together.

＊由于厚度有种类型，我们决定在平衡每种厚度的颜色的同时连接它们了，并将7到10扭曲在一起。

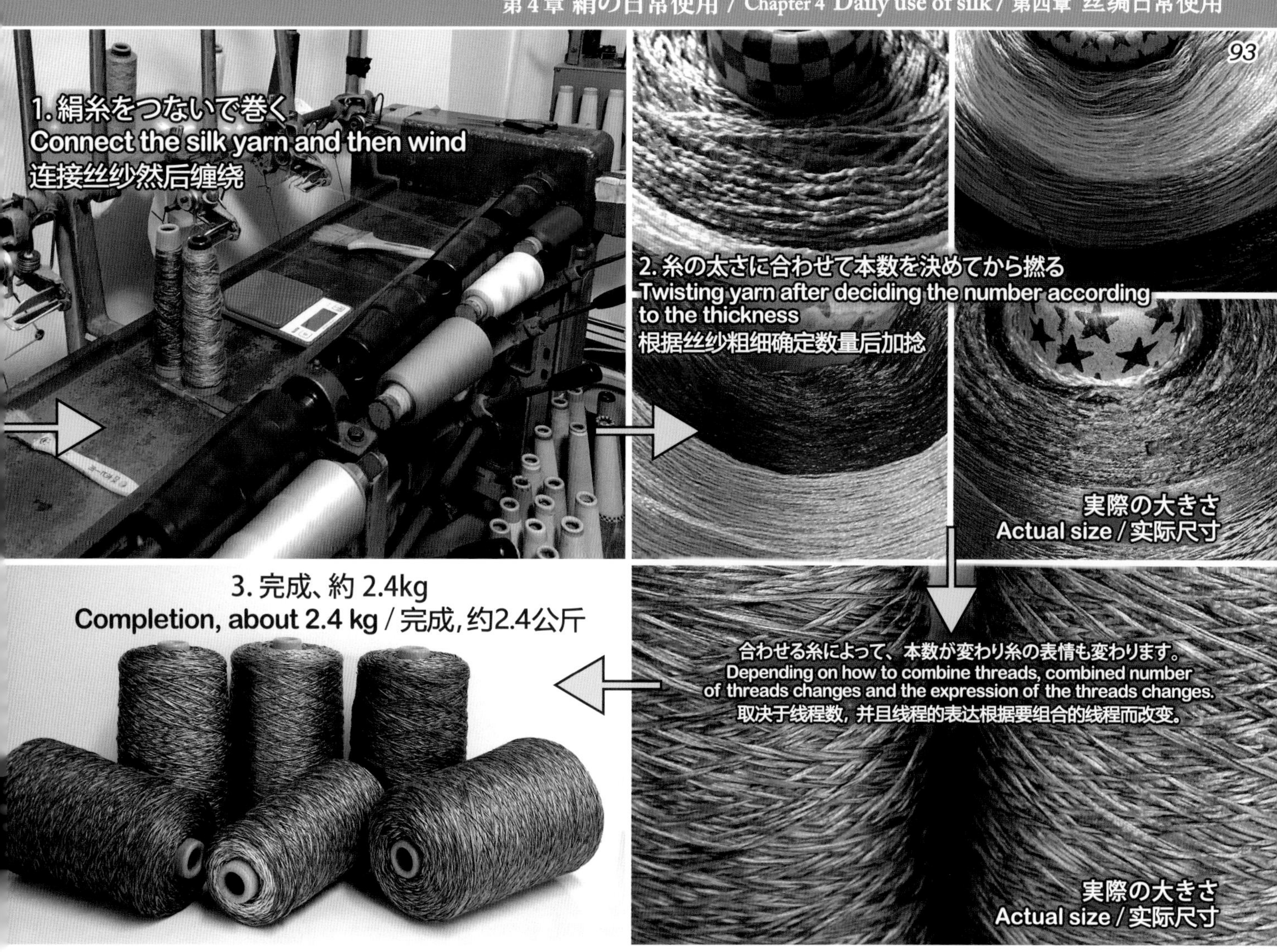

生糸が多く、糸の輝きが美しい

Lot of raw silk, so that beautifully shine / 大量的生丝，美丽的光泽

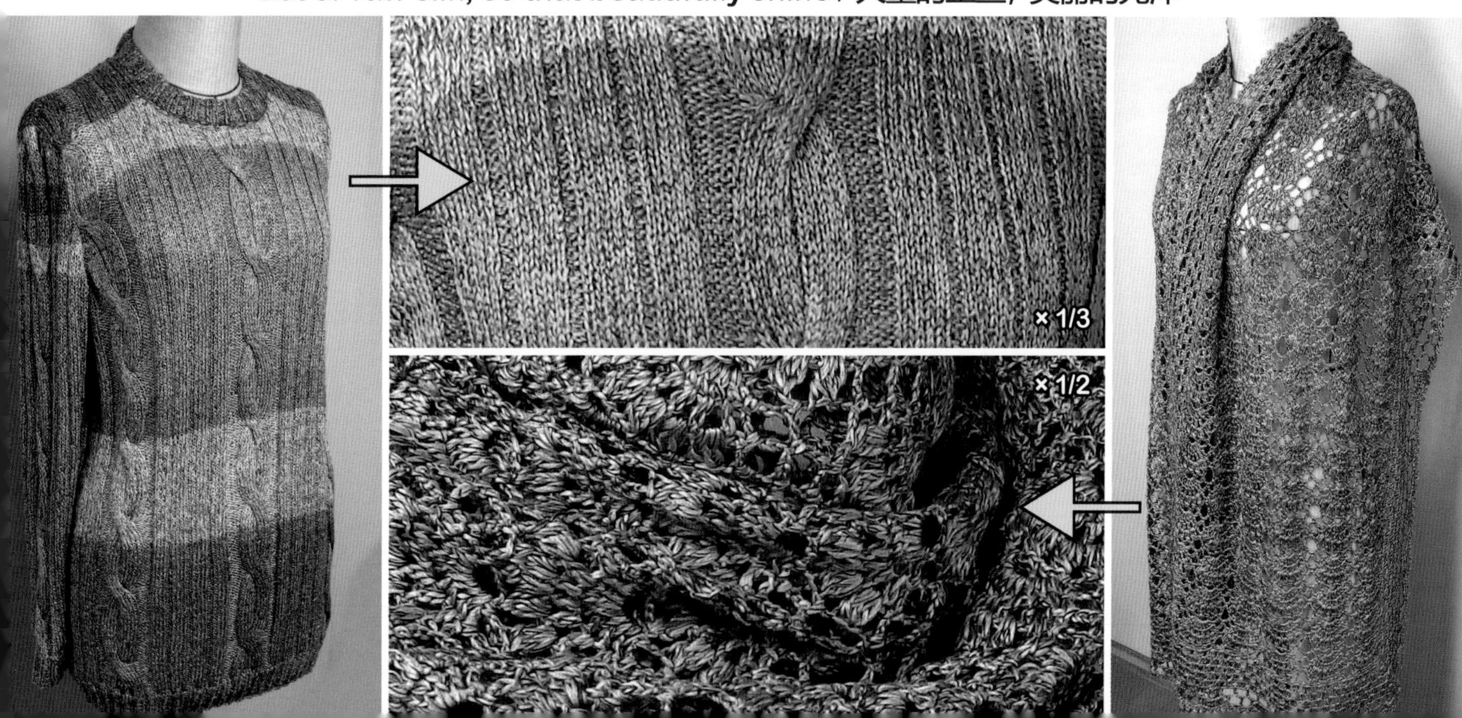

ファンシーヤーン（意匠撚糸）を編む
Knit a fancy yarn (design twisted yarn) / 我针织花式纱【设计加捻纱】

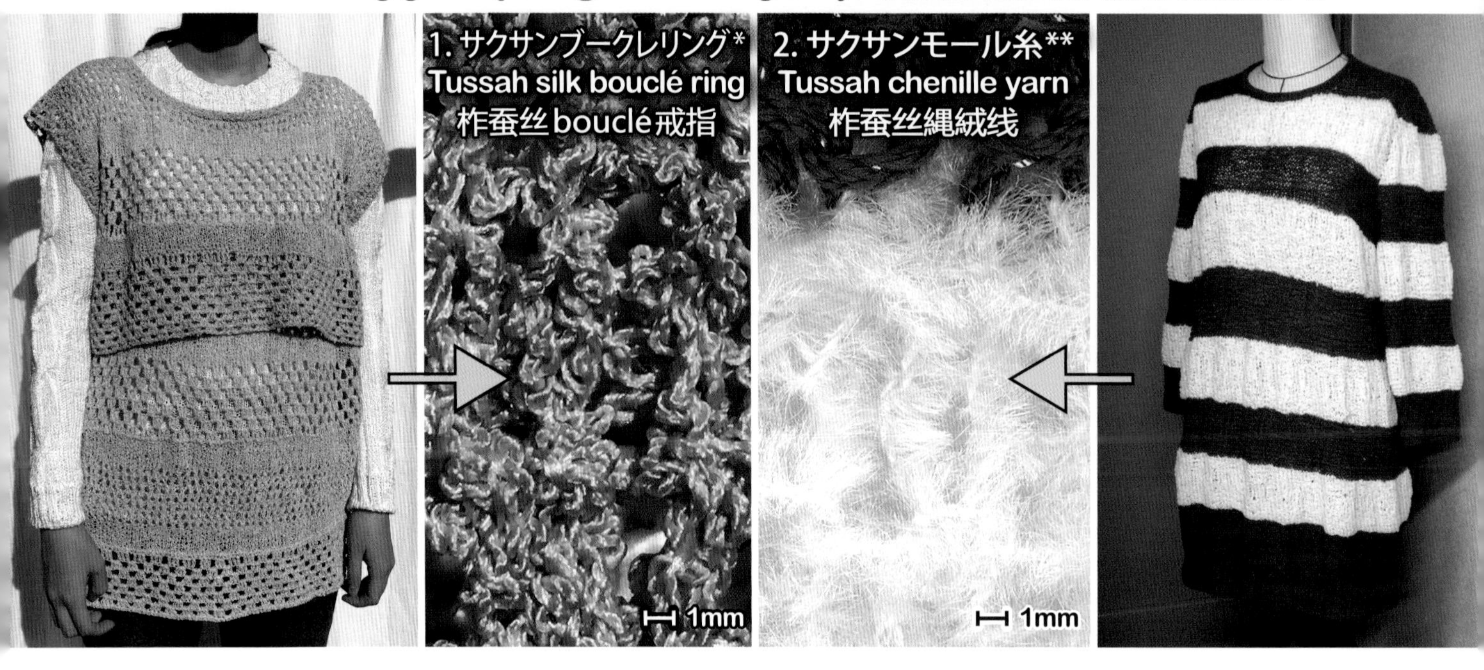

飾り糸の役割 / Role of decorative thread / 装饰线的作用

織物や編物に形状変化をつけるため、撚糸工程で太さのちがう糸を組みあわせたり、撚りの数や方向のちがう糸を組みあわせたりします。

太さのわりに柔らかな風合いや、節のある糸、粒のような飾りが所々に現れる糸、ループのある糸、特殊な色の組み合わせで効果を出した糸など、様々に工夫されています。

左上の画像のシャリ感を持つブークレリングは、軸の細糸に太糸を螺旋状に絡ませ、間隔を狭く押しつけて巻くことで、余った太糸がリング状に浮いてギザギザした感じになった糸です。仕上げに細糸を巻いて、短繊維が抜け出すのを防いでいるので、編む時に糸割れしにくく、完成後は毛羽立ちしにくい糸になります。

右上の画像のモール糸は、軽く、肌ざわりのよい糸です。カットした糸を軸糸に挟み込んで巻き付け、ビロード糸やシェニール（フランス語で毛虫）とも呼ばれます。糸が抜けやすいと聞きますが５年着用後、糸抜けを感じません。

In order to change the shape of woven and knitted fabrics, we combine yarns with different thicknesses in the twisting process, or combine yarns with different number and direction. It has been devised in various ways, such as a soft texture for the thickness, threads with knots, threads that show grain like decorations in some places, threads with loops, and threads that are effective with a combination of special colors.

Bouclé ring with the crispness of the image on the upper left is a thread with a jagged feel that the excess thread floats in a ring shape by winding thick thread spirally around the thin thread that is the axis and pressing it with a narrow gap. The fine yarn wound in the finish prevents the short fibers from coming off, the yarn is less likely to crack when knitted by hand, and the yarn is less likely to fluff after completion.

Chenill yarn in the upper right image is a thread that is light and good in touch for wearer. The cut thread is sandwiched between the axoneme and wrapped around it, and is also called velvet thread or chenille (caterpillar in French).

It is said that the thread will come off easily, but even after wearing it for 5 years, the thread has never come off.

为了改变机织物和针织物的形状，我们在加捻过程中结合了不同粗细的纱线，或者结合了不同数量和方向的纱线。它的设计方法多种多样，例如厚度柔软的纹理，带有打结的线纱，在某些地方显示颗粒状装饰的线纱，带环的线以及具有特殊颜色组合效果的线纱。

左上方图像脆的bouclé环是锯齿状的线，通过将粗线螺旋缠绕在细轴上并紧紧按压，使特粗线以环状漂浮。细线缠绕在精加工周围，以防止单纤维掉落，手工编织时纱线不会分裂，完成后不会起毛雛的。

右上方图像縄絨线，轻盈且在皮肤上感覚良好的纱线。是从切来的缠绕在轴线周围的线中切下并夹在的线，也称为chenille（法语毛虫）或天鹅绒线。

据说线很容易脱落，即使穿载了5年，线也没有感觉到任何脱落。

＊糸、＊＊糸：アトリエ トレビ P.134.　＊Thread, ＊＊ Thread: Atelier Trevi P.134.　＊线程, ＊＊线程: Atelier特雷维, 第134页。

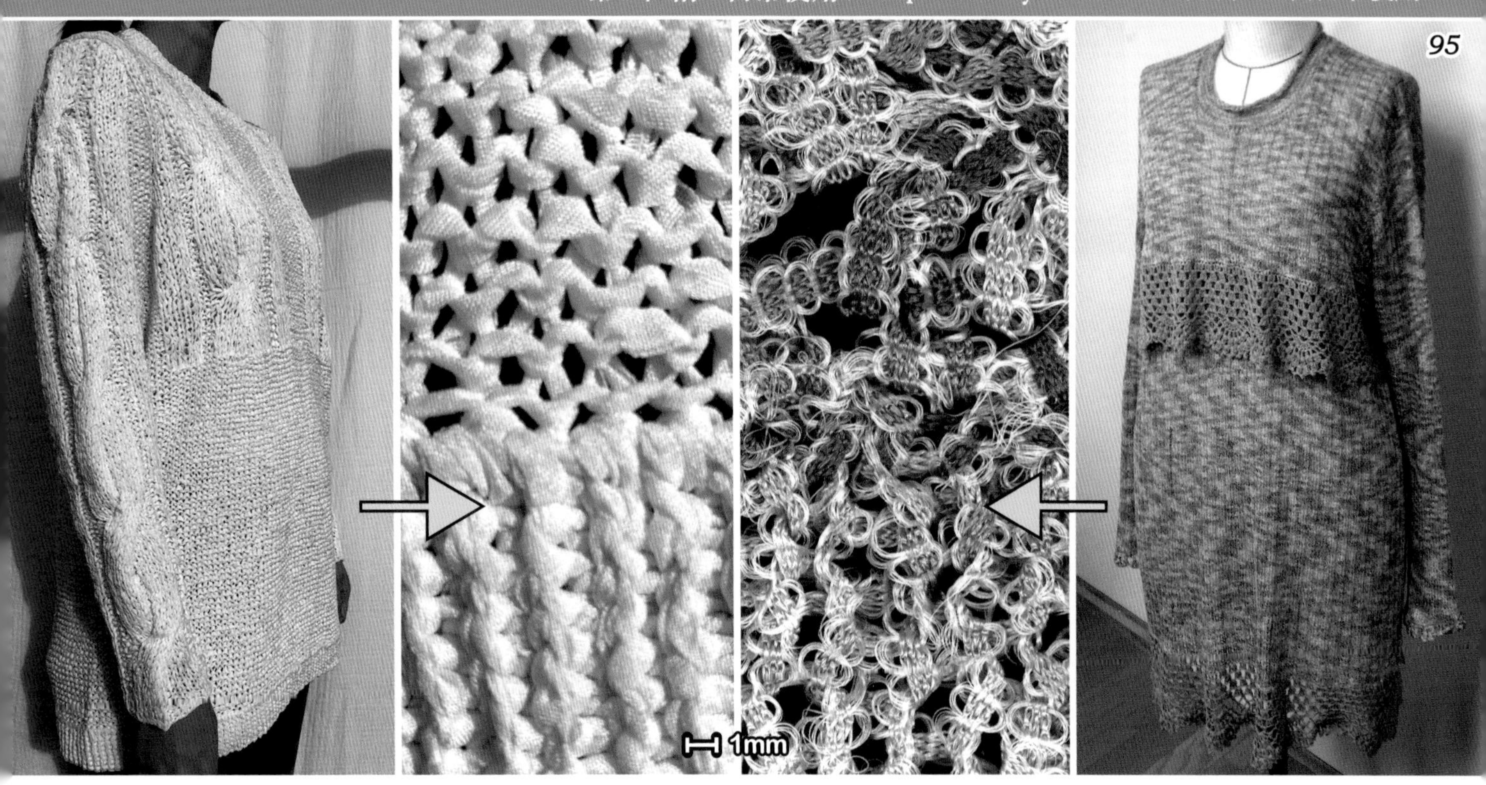

シルクリボンのニット / Silk ribbon yarn knit / 丝缎带线针织

シルクリボンは、手に入りにくいので出典は控えます。布を編む面白さとしては、裂き布がおすすめです。下のように厚い布よりも、スカーフくらいの薄地が編みやすいと思います。

Silk ribbon yarn are hard to come by, so I will refrain from introducing its sources. Cloth strips yarn is also recommended for the fun of knitting the cloth. I think a scarf-like thin fabric is easier to knit than a thick cloth as shown below.

丝缎带线很难获得，所以我将避免介绍其来源。作为布来编的针织的乐趣，建议使用布条线。我认为，比厚的织物更容易编织像围巾一样的薄织物，如下所示。

裂き布を編んだり織ったり / Knitting or weaving cloth strips yarn / 布条线的织和编

絹100%、絹+綿各50%、絹+リネン各50%の編み比べ
Knitting comparison of 100% silk, 50% silk + 50% cotton, 50% silk + 50% linen
100%真丝，50%真丝+50%棉，50%真丝+50%亚麻的针织造比较

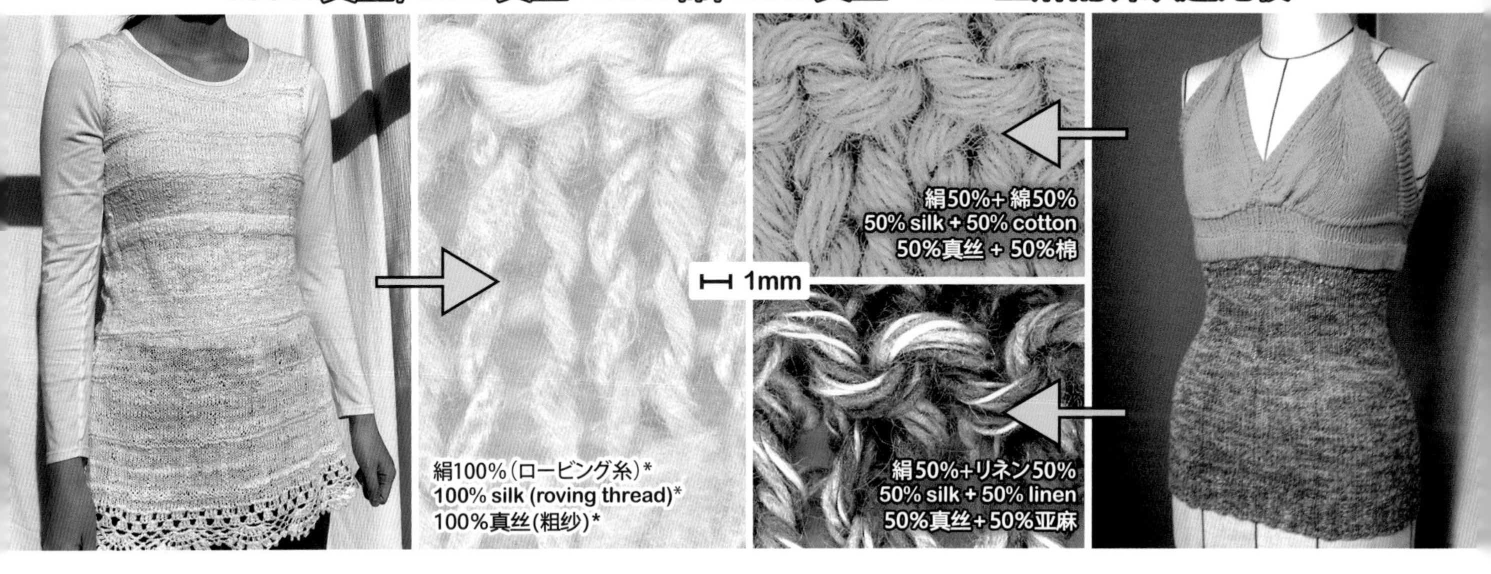

様々な絹糸で肌着を編む / Knit underwear with various silk yarns / 针织各种丝纱线内衣

シルクの普段使いには摩擦の弱さが問題になりますが、弱点以上の着心地のよさがあります。

最近の5年間は、手編みのシルク肌着を着用し、シルクの着心地の原点を見つめています。

1. すぐ乾き、吸湿、放湿する。

シルクの肌着は酷暑の中、体表に噴きだす汗を吸い込み、繊維に吸収して上着に染みだしにくくします。

また汗をかいてから冷房のきく場所に入った場合、繊維が含んだ汗は直接肌に当たらないので冷たくならず、いつの間にか乾きます。

2. アンモニア臭などの消臭効果。

真夏の汗や体臭が気になるときは、シルクの下着を着用すると安心です。

シルクのシーツや内掛け、枕カバー、パジャマは、介護で気になる、寝汗やアンモニア臭から介護者を開放する可能性があります。

Silk has weak point against friction and that is a matter for everyday use, but the wear comfort surpasses the weak point. For the last five years, I have been wearing hand-knitted silk underwear and looking at the origin of silk wear comfort.

1. Dry immediately, absorb moisture and release moisture.
During intense heat season, silk underwear inhales sweat that erupts on the surface of the body, absorbs it into the fibers, and prevents it from seeping into the outerwear. Also, if you enter a cool place after sweat, the sweat contained in silk fibers does not directly touch the skin, so it does not get cold, and sweat dries while you are unaware.

2. Deodorizing effect for ammonia, etc.
When you are worried about sweat or body odor during midsummer, wearing silk underwear makes you worry-free when visiting your workplace. Silk sheet, light blanket, pillowcase, pajamas, etc. may save caregivers from the smell of night sweat, ammonia, etc.

日常使用丝绸时摩擦是一个问题，但比其弱点以上穿着起来更舒适。在最近过去的五年中，我一直穿着手工针织的丝绸内衣，肌肤感到很舒服，这就是我参于丝织由原点。

1. 易干，会吸收水分并释放水分。
在炎热的季节，丝织内衣能吸收身体表面的汗液爆发，吸收到纤维中，并不会渗湿外衣。另外，如果出汗后进入凉爽的地方，纤维中的汗液不会直接接触皮肤，因此身体不会感到冷，并且会在不知不觉中变干。

2. 除臭效果，如氨味。
仲夏时，如果担心出汗或体臭，穿丝质内衣可使您感到放心。使用丝绸床单，轻毯，枕套，睡衣等可能会使身体摆脱盗汗和氨气的气味。因此，它有助于照顾者。

---

*スイス製、細いのにかさが出て編みやすい糸。同じ品質の糸は、残念ながら日本では作れないとのこと（現地購入品）。

*Made in Switzerland, it is a thin thread, but when it is knitted, it gives a volume and is easy to knit. Unfortunately, the same quality thread cannot be made in Japan (Purchased locally)

*瑞士制造的，它是细线，但是在针织时，它具有一定的体积并且易于针织。不幸的是，在日本无法制造相同质量的线（在本地购买）。

ネットロウシルク*
Net Raw Silk 1000 Denier*
网生丝1000旦尼尔*

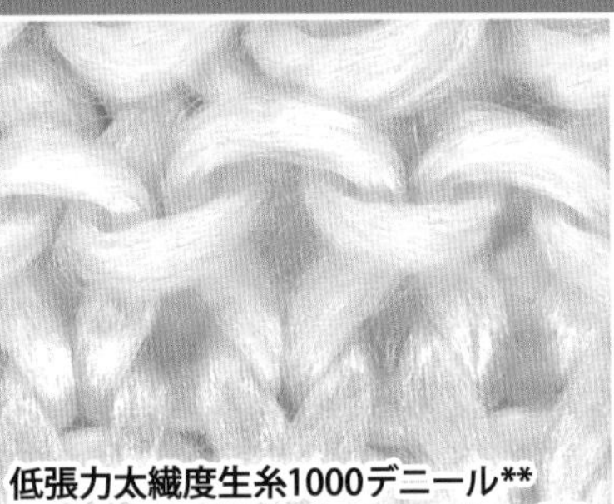
低張力太繊度生糸1000デニール**
Low-tension High-fineness Raw Silk 1000 Denier**
低张力高细度生丝1000丹尼尔**

絹紡紬糸***
Noil silk yarn***
紬丝 ***

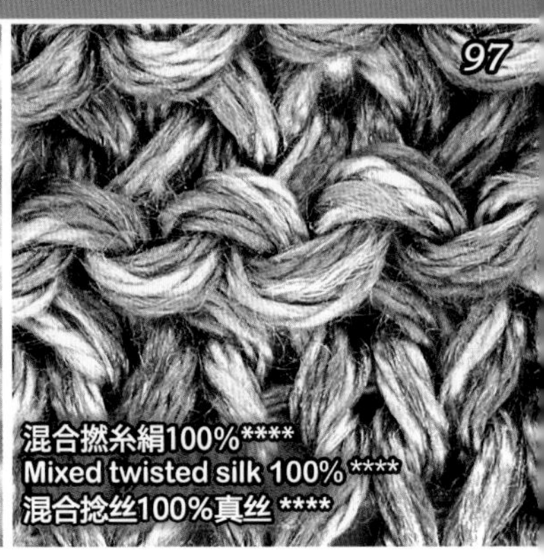
混合撚糸絹100%****
Mixed twisted silk 100% ****
混合捻丝100%真丝 ****

3. 人の肌に近いなめらかな肌触り。

するする滑るので編みにくい生糸が一番滑らかです。しかし、冬季には着た瞬間冷っとします。すぐ温まりますが、嫌う人もいます。

私は野蚕糸と絹紡糸を編んできましたが、最近は、生糸やほかの繊維との混紡糸を試しています。

4. 洗濯機で洗う場合。

洗濯袋に入れて、ほかの衣類と共に洗濯機で洗いますが、3 年くらいは補修の必要がなく、その後は、ほころびを直して着続けています。

3. Smooth touch that is close to human skin.

The raw silk that is hard to knit because it slides smoothly is the smoothest. However, I feel cold the moment I wear it in winter. It warm up quickly, but some people dislike it.

I have been knitting Wild silk yarn and spun silk yarn, but recently try raw silk yarn and blended yarns with other fibers.

4. When washing in the washing machine.

I put it in a laundry bag and wash it with other clothes in the washing machine, but there was no need to repair it for about 3 years, and after that, I repaired the open seam and continue to wear.

3. 接近人体皮肤的光滑触感。

由于光滑而难以针织的生丝是最光滑的。但是，我冬天穿的那一刻会感到冷。它很快变暖，但是有些人不喜欢它。我一直在针织野蚕丝纱和绢丝纱，最近又尝试使用生丝纱和其他纤维的混纺纱。

4. 在洗衣机洗涤时。

我将其放在洗衣袋中，放进洗衣机和其他衣物一起洗涤，这样大约 3 年以来都无需修理，此后，我修理绽线的并继续穿戴。

異種繊維混合撚糸 / Mixed and twisted different kinds of yarn / 各种纤维混合加捻纱

肌に密着して頻繁に洗濯する肌着は、絹繊維の着心地を比較しやすいので、ほかの繊維との混合撚糸を編みました。

綿との混紡糸は、半世紀以上前から内外に様々な糸があって馴染んできましたが、麻との混合撚糸は初めて編みました。左ページの拡大画像のように、糸の毛羽立ちが目立たない編み上がりになり、着心地がよいので驚きました。

機械編みのニットは、糸の弱点を整理してから機械編みされています。しかしさまざまな糸を収集し、手編みをする場合は、糸の弱点がでやすいので、日常的に着用して観察しています。

2019年春から「異種繊維混合捻糸」に挑戦し、セーターや肌着にしました。今後長く着用実験を続けて、身近な製品の製造につなげたいと思います。

For underwear that adheres to the skin and is washed frequently, it is easy to compare the comfort of the silk fibers, so I knit of mix yarn types by twisting thread with other fibers. I have been familiar with mixed spinning of silk and cotton, because its various yarns exist from more than half a century ago inside and outside of Japan.
But I knitted twisted yarn of combining silk and linen for the first time. As the magnified image on the left page shows, the fuzzing of the silk yarn is not noticeable and its wear comfort surprised me.

Machine-knit knit products are machine-knitted after the weaknesses of the yarn are sorted out. However, when various threads are collected, weak points are likely to occur due to hand-knitting, so I wear the knitted clothes on a daily basis and observe them. I knitted sweaters and underwear with the "heterogeneous fiber mixed yarn" that I have been challenging since spring 2019.

由于很容高度附着在皮肤上且经常洗的内衣中纤维的舒适性比较，因此我们还针织了不同纤维的混纺纱线。丝绸和棉的混纺在半个多世纪以来在日本和海外已为人们所熟悉，但我第一次针织丝绸和亚麻的混纺。如左页放大图所示，纱线的起毛并不明显，而且穿着舒适，所以我很惊讶。机针织针织的产品是在纱线的弱点被清除后进行机针织的。但是，我搜索并收集的线的手工编针织的针织也弱点很容易出现，因此我们每天都穿着它们进行观察。

自2019年春季以来，我们一直在挑战，用”各种纤维混合加捻纱”编织毛衣和内衣。我希望通过用长期的穿戴实验，制造出更多的貼身产品。

*糸、**糸、***糸：工房 風花
****糸及び、前ページの綿シルク、麻シルク：試作、WILD SILK MUSEAM

*Thread, **Thread, ***Thread: Atelier Kazahana
****Thread, Silk Cotton and Silk Linen on the previous page: prototype, WILD SILK MUSEAM.

*线程,**线程,***线程:工房 Kazahana
****前一页的线，棉丝和亚麻丝: 原型，野生丝绸博物馆

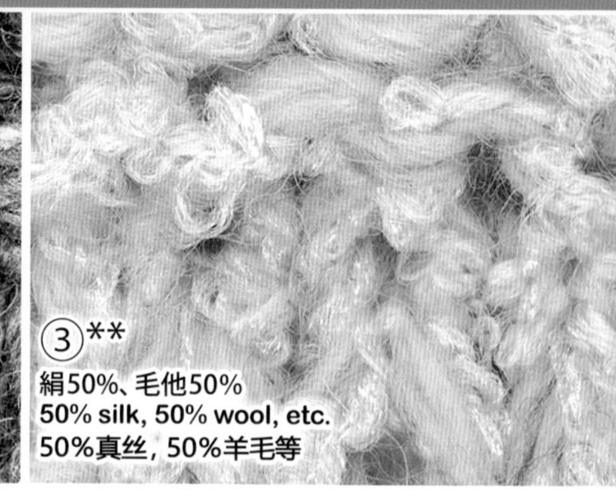

編み糸としての絹の特徴 / Characteristics of silk as a knitting yarn / 丝绸作为针织纱的特性

真綿は広げると、どんどん広がりますが握ると小さな塊になります。絹の糸は上の写真の①と②のように、1 本取りと、2本取りのゲージをほぼ同様に編むことができます。

①はゆるく編んだわけではありません。②は①を編んだ針と同じ針でかなりきつく編みましたが、きつく詰まった編み目には見えないでしょう。このように絹は、糸のふくらみ分がしぼみやすく、その分編み目がゆるむので、セーターの見頃丈や袖丈が伸びるのです。

麻や毛のように毛羽立つ糸と絹をより合わせた撚糸は、合わせた繊維のふくらみの部分に絹が入り込むように絡んで、絹糸だけが浮くことなく、とても編みやすくなります。そして 96ページの麻と絹、上の写真③の毛と絹の混合撚糸は、写真でも顕著なように、絹の単繊維が飛びださず、毛玉ができにくいようです。

If you spread the floss silk, it will spread more and more, but if you grasp it, it will become a small lump. As shown in ① and ② in the above picture, the silk yarn can be knitted almost in the same way both gauge at one strand and at two-strands. ① is not loosely knit. ② has been knitted tightly with the same needle as knitted in ①, but it will not be visible as the tightly knitted stitch. In this way, silk easily deflates thc bulge of the thread and so loosens the knitted stitches, which causes the increase of body length and sleeve length of the sweater.

Twisted yarn, which is a combination of yarn such as linen and wool with silk yarn is very easy to knit, because silk fiber tangles with counterpart yarn fiber as if silk fiber enters its bulge part and so silk yarn is not separated. And as is apparent in the photo of linen and silk on page 96, and the mixed twisted yarn of wool and silk in photo ③ above, the silk's single fiber does not appear and fiber ball is not made easily.

如果传播真丝綿，它将传播的越来越多了，但是，如果当被挤压时，它将变得很小块。如上图中的①和②所示丝线于，可以与单股和两股线相同的纱线規针编。①的，我没有织得較宽松。②与①都用相同的相同的编织针进行相当紧密的编织，但②，它看起来不像织得很紧。这样，丝绸很容易使线的隆起变松并松开针迹，由于瘪线而使线迹松散，并延长毛衣的长度和袖长。

将通过蓬松纱的纱线如亚麻或羊毛和丝绸纱线制成的加捻纱线，丝绸缠绕在一起，使其进入另一种纤维的膨满和起毛部分，因而丝纤维不突出，非常容易针织。而且如第 96 页上的亚麻和丝以及上面照片③中的毛和丝的混纺纱所示，丝的单丝纤维不会突出，也不容易出现纤维球。

糸の太さや編み方を変えて同じ糸を編み、着用試験をする
Knit the same thread by changing its thickness and its knitting method and perform a wearing test
通过改变线的粗细和针织方法来针织同一种线，并进行穿戴测试

スパッツを編む / Knitting spats / 手工编緊身褲

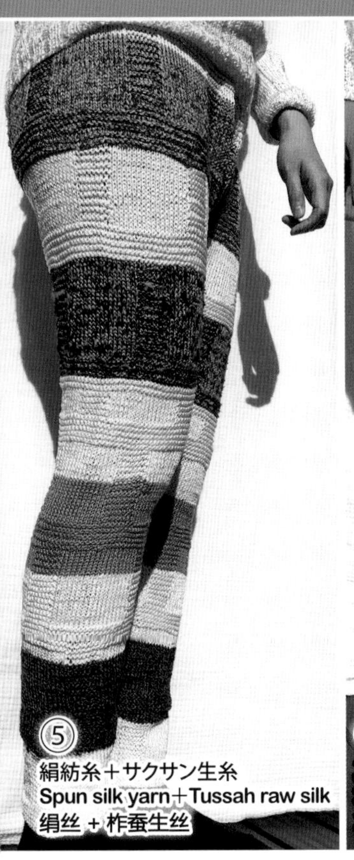

⑤
絹紡糸＋サクサン生糸
Spun silk yarn＋Tussah raw silk
绢丝 + 柞蚕生丝

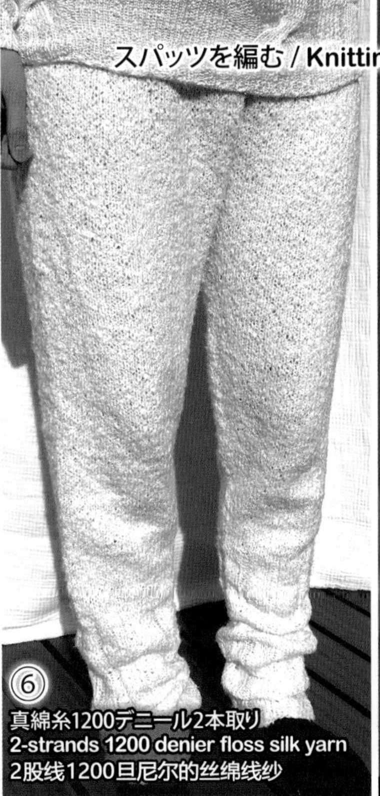

⑥
真綿糸1200デニール2本取り
2-strands 1200 denier floss silk yarn
2股线1200旦尼尔的丝绵线纱

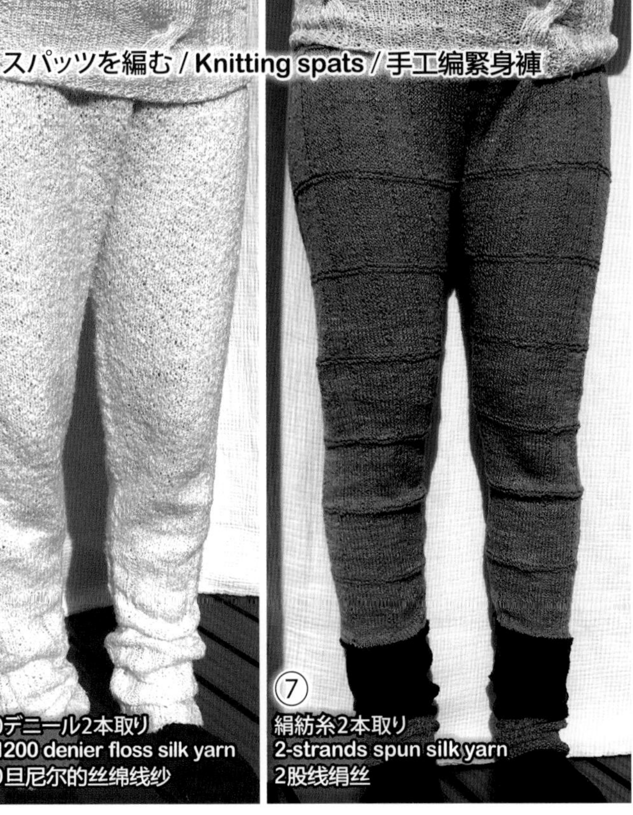

⑦
絹紡糸2本取り
2-strands spun silk yarn
2股线绢丝

⑧***
特殊染色絹紡糸2本取り
2-strands special dyed spun silk yarn
2股线特殊染色的绢丝

運動領域の多い膝と腰を包むスパッツは、糸の劣化時期がわかりやすいので、様々な糸を編んでいます。

1950 年代半ばころから家庭用編み機の普及とともに、美しい色の毛糸が豊富に販売され、各家庭でセーターやズボン下、毛糸のパンツが編まれるようになりました。

上のスパッツは、その時代から長く機械編みで編まれていた形をアレンジしました。

⑤と⑧は足首を細く、⑥と⑦は丈長にして足首にゆとりをつくり、レッグウオーマー状にしました。

Spats wrapping around the knees and waist, where move frequently with body movement, are knitted with various threads because it is easy to see when the threads deteriorate.

With the prevalence of household knitting machines since mid-1950s, a large number of beautifully colored wool yarns have been sold, and sweaters, leggings and Woolen pants have been knitted at each home. The design of spats in the picture above was modified from the design that had long been knitted by machine. ⑤ and ⑧ have thin ankles, and ⑥ and ⑦ were knitted to make length long and ankle part ample, that is, like leg warmer.

包裹膝盖和臀部，活动多的运緊身褲，因为很容易看到线何时变质，所以用各种线针织而成。

1950 年代中期以来，随着家用编织机的广泛使用，出售了许多色彩艳丽的羊毛纱，每个家庭都编织了毛衣，緊身褲和羊毛裤衩。上图中的緊身褲是根据自那时以来长期通过编织机而成的形状设计的。⑤和⑧的脚踝较薄，而⑥和⑦的较长且脚踝较厚，其设计成看起来像暖腿套。

機能的な使い方 / Functional usage / 功能性的用法

ウールのようにチクチクしない絹は、薄い繊維の上に羽織れ、1 年中着られる便利な繊維です。

肩から滑り落ちるショールは、家事には不向きなので、左ページ④のようなショートベストが重宝します。ショールほど糸を使わないので、余り糸を利用して短時間で編めます。

Silk, which doesn't make feel scratchy like wool, is a convenient fiber that can be worn all year round because it can be worn over thin clothes. Shawls that slide off your shoulders are not suitable for housework, so the short vest on the left page ④ is useful. Since it does not require as much yarn as a shawl, it can be knitted in a short time using surplus yarn.

不会像羊毛一样刺痛皮肤的丝，是一种方便的纤维，全年穿着，因为它可以穿在薄衣服上。

从肩上滑落的披肩不适合做家务，因此左页侧第④所示的短背心很有用。由于它使用的纱线不如披肩那么多，因此可以用較短时间，使用余剩纱线进行编织。

---

*糸, **糸：試作、WILD SILK MUSEAM。
***糸：東京アートセンター

*Thread, **Thread: prototype, WILD SILK MUSEAM.
***Thread: Tokyo Art Center

*线程, **线程：WILD SILK MUSEAM (野生蚕丝博物馆)。
***线程:Tokyo Art Center (东京艺术中心)

余り布、余り糸で便利グッズを
Make useful items with surplus cloth or thread
用余剩布和线做有用的物品

着物のリユース / Kimono reuse / 和服再用

着物*の身頃布を肩で切り、縫い合わせて140×180cmくらいの大きな布を作ります。
裏地もはぎ合わせて袋状に縫い、表に返して周囲にミシンステッチをかけます。布をはぎ合わせた場所にもステッチをかけると、洗濯に強くなります。

Cut the body cloth of the kimono at the shoulders and sew it together to make a large cloth of about 140 x 180 cm. The lining is also sewn together and then into a bag shape, and after turning it over, the sewing stitch is applied to the periphery. Apply sewing machine stitches to the seams where the cloth is together, put the sewing the cloth align the front and back of together, it will be stronger against washing.

用肩膀剪开和服，并将其缝在一起，制成约140x180厘米的大布。将衬里缝在一起，缝成袋状，返回表部后，在其外周周围应用缝纫机针迹。布料的正面和背面缝合时，如果缝纫机针迹缝在一起的布接縫上，将很耐洗。

肩掛け、ショール、ひざ掛け、毛布などの内掛けに
Shoulder wrap, shawl, lap blanket and for inner blanket
用于長披布，披肩，盖膝套和毛毯的内側。

直線断ちの服 / Straight cut clothes / 直线裁剪的衣服

3枚のショールで作った上着
Outerwear made of 3 shawls
三件披肩制成的夹克

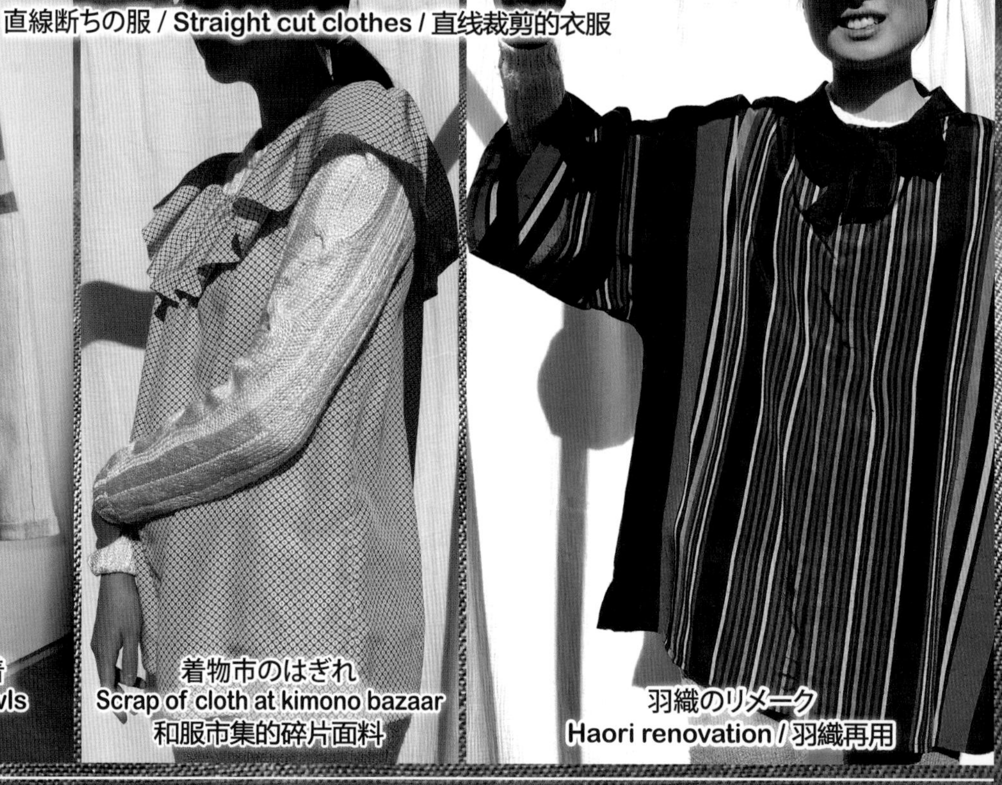

着物市のはぎれ
Scrap of cloth at kimono bazaar
和服市集的碎片面料

羽織のリメーク
Haori renovation / 羽織再用

*染色：㈱みなみ紬、ワークショップにて。
①～⑥糸：工房 風花、⑦糸：アトリエ トレビ。
⑧糸：WILD SILK MUSEUM。

* Dyeing: At the Minami Tsumugi, Workshop.
①～⑥ Thread: Atelier Kazahana.
⑦ Thread: Atelier Trevi.
⑧ Thread: WILD SILK MUSEUM.

*染色：在 Minami Tsumugi的工作坊。
①～⑥线程：Kazahana工房。
⑦线程：Atelier 特雷维。
⑧线程：野生丝绸博物馆。

シルクとウールの長所と短所 / Advantages and disadvantages of silk and wool / 丝绸和羊毛的优缺点

シルクは毛羽立ちますが、ウールほど密に単繊維が絡み合わないので、バルキーセーターでも風を通します。ウールは風を通しにくいので、冬季、バルキーセーターを着て戸外で作業ができます。しかし作業をして汗ばむと、湿気を逃がしにくいので汗が体表に残り、身体を冷やします。

一方、シルクは歴史上、美しい面ばかりが重要視されてきました。特に靴下や肌着に使う心地よさは、為政者や一部の特権階級のみが知ることであり、多くの人々は知る由もありませんでした。

シルクが一般的に販売されるようになってから百数十年、シルクの機能性が肌着などに利用されるようになって数十年。残念ながら養蚕業が世界的に失われつつあります。

Silk is fuzzy, but since the monofilaments do not entangle as closely as wool, silk bulky sweater allows draft. Wool is hard to allow draft, so in winter you can wear a bulky sweater and work outdoors. However, when you work and sweat, it is difficult for moisture to escape and sweat remains on the body surface and cools the body.

On the other hand, only the beautiful aspect of silk has been emphasized in history. The comfort of socks and underwear, in particular, was known only to the administrators and some privileged classes, and many people could not know it. Silk began to be sold to the average people only one hundred and a few decades ago and its usage to underwear for its excellent function started only a few decades ago. Unfortunately, the sericulture industry is being lost worldwide.

丝绸的蓬松和起毛，是由于单纤维的缠结不如羊毛緻密，即使制感大件毛衣，也較通风 。羊毛很难通风，此在冬天，您可以穿大件毛衣并在户外工作。但是，干活出汗时，潮气很难散失，汗液残留在身体表面，从而使身体受涼。

另一方面，在丝绸历史中，美丽的一面一直被认们所重視。尤其是袜子和内衣的舒适性只有为政者和一些特权阶层才知道，而许多人却不知的。丝绸被普遍出售以来已经过去了一百多年，而丝绸的功能性，被用于内衣才过去了数十年。遺憾的是，蚕桑业正在全球范围内流失。

シルクの日常着の耐久力 / Durability of silk's daily wear / 丝绸日常穿着的耐久性

連続着用実験：エリシルク平織藍染生地*
Continuous wear experiment: Eri silk plain weave indigo fabric*
连续穿着实验：樗蚕丝(Eri silk)平纹靛蓝染色织物*

4年4カ月日々6時間以上素肌に連続着用し、随時洗濯を繰り返した生地の経年変化
Change of fabric with the passage of time: wearing on bare skin, 6 hours or more daily for 4 years and 4 months and repeating wash when needed.
织物测试，四年四个月每天连续六小时或更长时间裸露皮肤连续穿着，并反复清洗

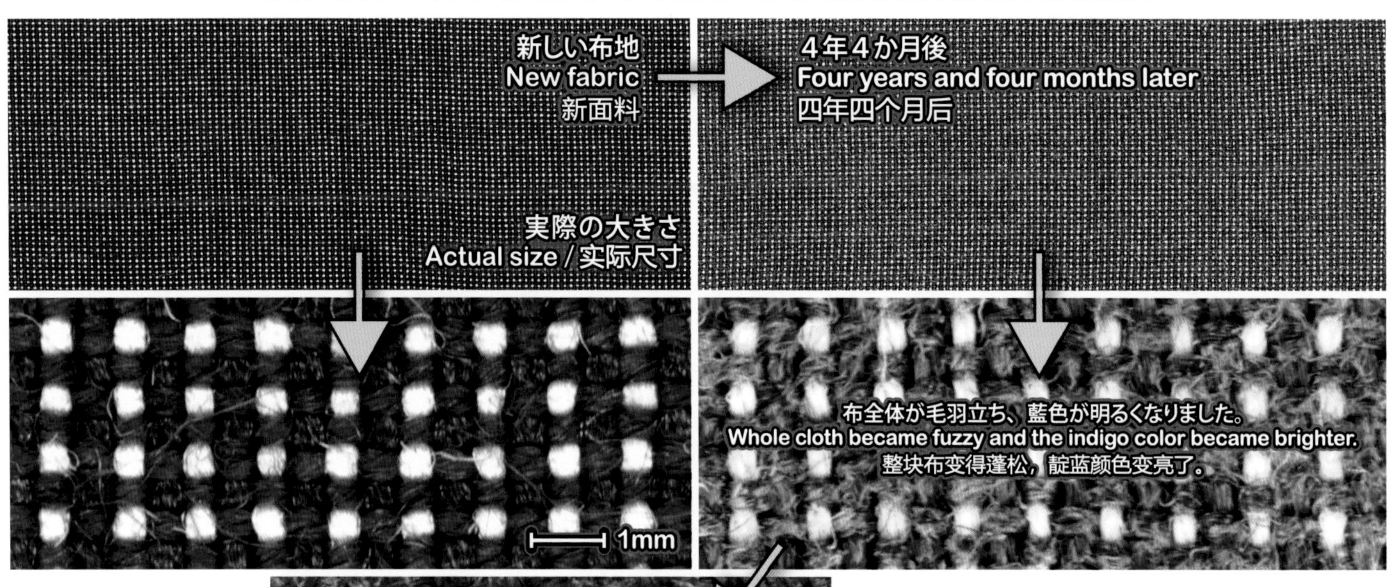

エリシルクは家蚕と同じくらいの繊度です。P.105の寝間着と一緒に着用しました。浴衣地くらいの厚さなので、朝洗濯をして干しておけば夕方には乾きました。

4年4か月、必要に応じて洗濯をし、毎晩着用し続けて生地の劣化を観察しました。

1．3年着用後、裾の折山の糸が切れました。
2．そのころに、ベルト裏のこすれた部分の生地が、薄くなりはじめました。
3．4年4か月後、薄くなった生地の緯糸が一気に千切れました。

以上の結果、藍染めの堅牢度に問題があるものの日常着としての強度は十分だと思いました。

Eri silk is the same level of fineness as that of the silkworm. I made pants and wore them with the nightwear on page 104 together. It's about the thickness of a yukata cloth, so I washed it in the morning and hung it, and then it got dry in the evening. 4 years and 4 months, I washed it as needed and continued to wear it every night to observe the deterioration of the fabric.

1. After wearing it for 3 years, the thread on the crease of the hem was cut.
2. At the same time, the fabric backing of the belt that was rubbed with the body began to become thinner.
3. After 4 years and 4 months, the weft yarn of the fabric that got thinner was cut at a stroke.

From the above result, I thought that that fabric is strong enough for everyday wear, though indigo dyeing has a problem with its fastness.

樗蚕丝绸具有和蚕一样的细度。我制作了裤子，并搭配第105页的睡衣，一起穿着。大约是浴衣的厚度，所以如果早上洗它，并挂起来，到晚上会干燥。四年四个月，我根据需要即其清洗，并每天晚上继续穿着以观察织物质地的变化。

1. 佩戴三年后，下摆的线断了。
2. 同时，皮带背面摩擦身体的织物开始变薄。
3. 四年四个月后，变薄的面团的纬纱突然破裂。

由于上述原因，我认为靛蓝染色的牢度有问题，但足以应付日常穿着。

＊エリシルク平織り、＊＊エリシルクデニム織：アトリエトレビ（P.134）。
＊＊＊㈱アート（P.138）：群馬県桐生市。

*Erisilk plain weave, **Erisilk denim weave: Atelier Trevi (P.134).
*** Art Co., Ltd., (P.138): Kiryu City, Gunma Prefecture.

＊樗蚕丝平纹编织，＊＊樗蚕丝粗斜棉布编织：Atelier 特雷维(P.134)。
＊＊＊Art Co., Ltd., (P.138)：群马县桐生市。

水洗いによる退色実験：藍染のエリシルク、デニム織り ** キトサンSP(シルクプロテイン)加工
Fading experiment by washing with water: Indigo dyed Eri silk, denim weave ** Chitosan SP (silk protein) processing
用水洗涤褪色实验：靛蓝染色的樗蚕丝绸，粗斜棉布梭织 ** 壳聚糖SP(丝蛋白)加工

加工：㈱アート、実験布提供及び水洗い：アトリエトレビ
Processing: Art Co., Ltd., Provide laboratory cloth and wash: Atelier Trevi / 流程：Art Co., Ltd.， 提供实验用布并用水冲洗：Atelier 特雷维

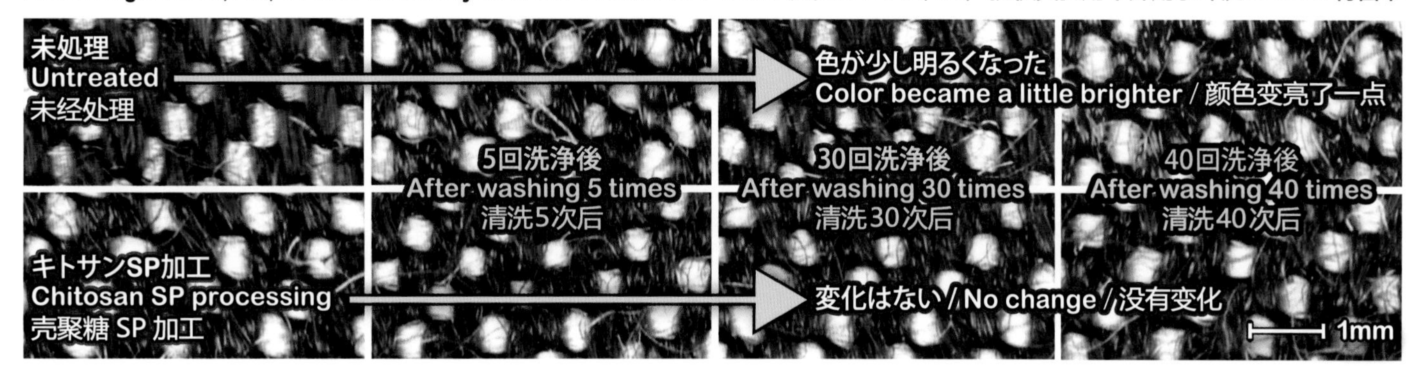

通常の色止め加工：水10リットルの中で布を50回押し、40回水を交換する
Normal color stop processing: Press the cloth 50 times in 10 liters of water and change the water 40 times
正常的颜色停止处理：在10升水中压布50次，然后换水40次

藍染め製品の多くは色落ちします。私は、左ページのズボンを裁断する前に2回水洗いをしました。しかし、アトリエトレビでは製品化後の色落ち防止のため、裁断前の水洗いで、水に色が出なくなるまで洗っています。

過去の経験から、SP加工が色止めにもなりうることを予想していた私は、アトリエトレビのエリシルクに本加工をすすめました。そして完成した布と、未処理の布で仕立てたズボンを、通常の水洗いをする手順で洗う実験を依頼しました。

Many indigo dyed products lose their color. I washed the trousers on the left page twice with water before cutting. However, at Atelier Trevi, clothes are washed with water until no color appears in the water before cutting the clothes in order to prevent color fading after commercialization. From my past experience, I expected that SP processing could also be effective for color fading stop, so I recommended this processing to Eri Silk cloth of Atelier Trevi. Then, I asked for an experiment to wash the finished cloth and the pants made of untreated cloth with normal washing procedure.

许多靛蓝染料会褪色。我在剪左页的裤子之前洗了两次面料。为防止商品化后变色，特雷维工房(Atelier Trevi)用水清洗布，直到在切剪开布之前在水中没有掉色出现。

根据我过去的经验，我予想SP处理也可能是一个防掉色处理，因此我推荐此处接特雷维工房的樗蚕丝绸来处理。然后，我进行一成个实验，以处理过的布和未经处理过布做成裤子用普通水洗程序来洗涤。

実験方法と結果 / Experimental method and result / 实验方法与结果

1．6枚の試験片を、処理布製と未処理布製のズボンの内側に縫いつけました。
2．水洗い5回ごとに試験片を外しました。
3．未処理布は30回まで水に色がでました。
4．40回終了後、未処理布はウエスト布が縫い目に沿って切れました(下の写真)。

アトリエ トレビでは退色以外に、縫製やアイロン作業中に色落ちや色移りがないことを確認しました。

本実験の報告後、㈱アート***は公的機関に、次ページの試験を依頼しました。

1. 6 test pieces were sewn on the inside of treated and untreated trousers.
2. Test piece was removed one by one after every 5 washes.
3. In case of untreated cloth, color appeared in water till 30 times washing.
4. After 40 times washing, in case of the untreated fabric the waist fabric was cut along the seam (picture below).

Atelier Trevi confirmed that the processed cloth did not fade or color transfer during sewing or ironing. After the report of this experiment, Art Co., Ltd.*** requested a test on the next page to a public institution.

1. 六个试件缝在了将处理过的和未处理的裤子内部。
2. 每清洗五次后一次后取下试件。
3. 未经处理的布料会连续掉色 30 次。
4. 40 次后，未处理的布的腰部布沿接缝裂开（下图）。

特雷维工房确认加工后的布在缝制或熨烫过程中没有掉色或颜色转移。在报告了该实验之后，Art Co., Ltd.*** 请求公共机构在下一页进行测试。

未処理 / Untreated / 未经处理　1mm

キトサンSP加工 / Chitosan SP processing / 壳聚糖SP加工

洗濯堅牢度、物性変化試験結果
Washing fastness, physical property change test result / 耐洗牢度及物性变化的试验结果

加工：㈱アート / 試験機関：群馬県繊維工業試験場
Processing: Art Co., Ltd., / Testing institution: Gunma Prefecture Textile Industry Test Site
加工：Art Co., Ltd. / 试験机关：群马县纺织工业试验场

試験項目 Test items / 测试项目	未処理 Untreated / 未经处理			キトサンSP加工 Chitosan SP processing / 壳聚糖SP加工		
洗濯試験 Laundry test 洗衣测试	変退色 Discoloration 褪色	汚染(綿) Pollution (cotton) 污染(棉花)	汚染(絹) Pollution (silk) 污染(丝绸)	変退色 Discoloration 褪色	汚染(綿) Pollution (cotton) 污染(棉花)	汚染(絹) Pollution (silk) 污染(丝绸)
	4-5級 Level 4-5 / 4-5级	3-4 級 Level 3-4 / 3-4级	3-4 級 Level 3-4 / 3-4级	4-5級 Level 4-5 / 4-5级	4級 Level 4 / 4级	4-5級 Level 4-5 / 4-5级
摩擦試験 Friction test 摩擦试验	たて / よこ方向 Vertical / Horizontal direction 垂直 / 水平方向	たて / よこ方向 Vertical / Horizontal direction 垂直 / 水平方向		たて / よこ方向 Vertical / Horizontal direction 垂直 / 水平方向	たて / よこ方向 Vertical / Horizontal direction 垂直 / 水平方向	
	乾燥2級 Dry Grade 2 / 干燥2级	湿潤1級 Wet grade 1 / 湿润1级		乾燥3級 Dry Grade 3 / 干燥3级	湿潤1級 Wet grade 1 / 湿润1级	
引裂き強さ試験 Tear strength test 撕裂强度测试	たて方向 Vertical direction / 垂直方向	よこ方向 Horizontal direction / 水平方向		たて方向 Vertical direction / 垂直方向	よこ方向 Horizontal direction / 水平方向	
	17.7 N	21.1 N		24.5 N	25.6 N	
磨耗強さ試験 Abrasion strength test 耐磨强度测试	179回 / 179 times / 179次			265回／ 265 times / 265次		

実験及び試験結果から考えたこと / Discussion / 来自实验和测试结果的思考

SP加工は、綿や化学繊維に絹の特性を付与するために考案された加工なので、おそらく絹への加工はテーマとされなかったでしょう。

製品化する企業の実験で、本加工に色止めと生地強度を上げる効果が見出されました。

一方で上記試験の結果、絹繊維の弱点である洗濯、摩擦、引き裂き、磨耗の項目で数値の改善が認められました。

したがって絹繊維のSP加工は、「水洗いをして気楽に着られる絹の日常着」の創出の可能性を持つと思いました。

綿繊維のSP加工は、絹と綿布の特性を兼ね備えた繊維として各所で製品化されています。

私は、㈱アートが加工した綿繊維に興味を持ち、加工方法などを見学後提供された布＊で寝間着を作り、P.102のエリシルクのズボンと共に、次頁の着用実験をしました。

Since SP processing was devised to impart silk characteristics to cotton and chemical fibers, processing into silk probably has not been the theme.

In an experiment conducted by a commercializing company, it was found that this processing had the effect of increasing the color fastness and the fabric strength. On the other hand, as a result of the above test, the improvement shown as numerical number was recognized in the items of washing, friction, tearing, and abrasion, which are the weak points of silk fiber. Therefore, I thought that SP processing of silk fibers has the potential to create " silk for everyday wear that can be washed and worn at ease."

Cotton fiber after the SP processing is commercialized in various places as a fiber that has the characteristics of silk and cotton cloth.

I was interested in cotton fibers processed by Art Co., Ltd., and after observing the processing method, I made a sleepwear with the cloth* provided by the company, and wore it with the Eri silk slacks on page 102 together, and I did wearing experiment of the next page.

由于设计SP处理是为了赋予棉和化纤，以丝绸特性，所以加工成丝绸可能是计划外的。在将该产品商业化的公司进行的实验中，发现该处理具有提高防掉色和织物牢度的效果。另一方面，作为上述测试的结果，在洗涤，摩擦，撕裂和磨损这些是丝绸纤维的弱点的项目中发现了数值上的改进。因此，我认为丝纤维的SP加工，有可能创造出【可以用水洗涤，日常可舒适穿着的丝绸】。

而今棉纤维的SP加工以使它成为一种兼备丝和棉特性的纤维，在各地都已商品化。我对Art Co.,Ltd.处理的棉纤维感兴趣，在参观处理方法等后，用公司提供的面料＊做睡衣，和搭配第102页樗蚕丝绸裤子，我进行了次页的穿着实验。

＊＊提供された布：2色各140㎝幅×2m

＊Provided cloth: 2 colors each 140 cm width x 2 m

＊提供的布料：
每种140厘米宽x 2m两种颜色

キトサンSP加工をしたコットンの布による、長期着用試験報告
Long-term wearing test report with cotton cloth treated with chitosan SP
脱乙酰壳多糖SP处理的棉布的长期穿着测试报告

開発会社：㈱アート / 試験機関：群馬県繊維工業試験場
Development company: Art Co., Ltd. / Testing institution: Gunma Prefecture Textile Industry Test Site
开发公司: Art Co.,Ltd. / 试験机关: 群马县纺织工业试验场

㈱アートでは、繭毛羽から抽出したSP（セリシン・フィブロイン各50％混合液）を、キトサンのプラス電荷とシルクのマイナス電荷のイオン結合を利用して繊維に定着しています。

私は、SP加工をした衣服を、洗濯しながら着用した場合、何年効果を保つか興味を持ちました。

過去のデータでは、洗濯50回でシルク及びキトサンの脱落の割合は約半分。しかしナノ銀を使用しているので、洗濯を100回行った後でも防臭・抗菌・消臭性能の効果は持続するとのことでした。

2014年4月、提供されたSP加工の綿のダブルガーゼで、寝間着(上の写真)2枚を作り、毎晩着用しました。

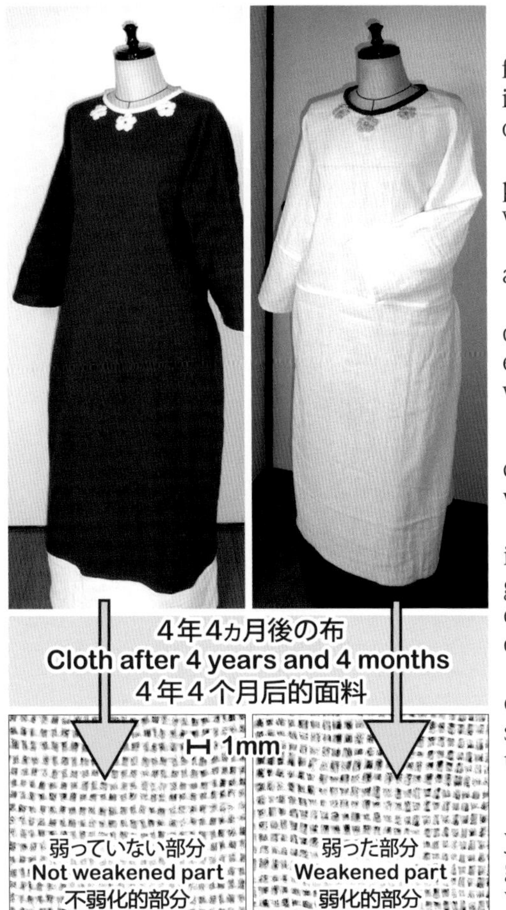

At Art Co., Ltd. SP (mixed liquid of sericin and fibroin, each 50%) extracted from floss of cocoon is fixed to fibers by ionic bondage using positive charge of chitosan and negative charge of silk.

I was interested in that how many years the SP processed clothes would keep their effect when wearing them while washing appropriately.

According to past data, the rate of drop of silk and chitosan is about half after 50 washes.

However, since nano-silver is used, the deodorant and antibacterial effects were explained to continue even after 100 times of washing.

In April 2014, using the SP processing double cotton gauze provided, we made the two sleeping wears (photo left) and wore them every night.

I often develop fever at night* and when I was in the high fever near 40 degrees, cotton wears got wet completely and I could not sleep without changing my clothes. Even if the silk gets wet, it dries more and more, so it doesn't wake up.

In the case of this SP processing, processed cotton gets a little moist but it didn't hinder sleep, and cotton double gauze gave good texture.

In August 2018, the fabric weakened after 4 years 4 months. Since the fabric part of the back got thin and was apt to be moist with sweat, the wearing experiment was ended.

Continued to the next page.

夜間に熱を出す*ことが多い私は、40度近い高熱時には木綿はぐっしょりぬれ、着替えずには眠れませんでした。

シルクは湿ってもどんどん乾くので目覚めません。本SP加工の場合は、少ししっとりしましたが目覚めることはなく、コットンダブルガーゼの風合いのよさがありました。

2018年8月、4年4か月を経過して布地が弱り、生地の薄くなった背中が汗ばむようになったので、着用実験を終了しました。次ページに続く。

在 Art Co., Ltd., 壳聚糖的正电荷和丝绸的负电荷结合的方法将从茧衣的绒毛提取的SP(50%的丝胶和丝素蛋白混合物)固定在纤维上。我对经过SP处理的衣服，洗涤，穿着，能保质多为年感兴趣。根据过去的数据，经过50次洗涤，丝绸和壳聚糖的掉落率约为一半。然而，由于使用了纳米银，即使经过100次洗涤，其除臭，抗菌和脱臭效果仍将持续。

2014年4月，我们使用提供的SP加工双层棉纱布，制作了上述两件睡衣(如上图)，并每天晚上穿着。

我经常在晚上发烧*，当我在40度附近发高烧时，棉质睡衣弄湿了，如果不换衣服我就无法入睡。

即使丝绸被弄湿了，因为很快会干，因此熟睡的我也不会醒来。在这种SP加工的织物，它有点汗湿但人不会醒来，显示了棉双层纱网良好的质地良好。

在4年4个月后，织物在2018年8月变弱化，织物变薄背中部的也发汗，因此穿着实验完成。接下页。

*夜間に熱を出す：幼少期から続く扁桃腺肥大で、高熱が出ます。

*Develop a fever at night: High fever due to tonsillar hypertrophy that continues from childhood.

*晚上散发热量：从童年开始，因扁桃体肥大引起的高烧。

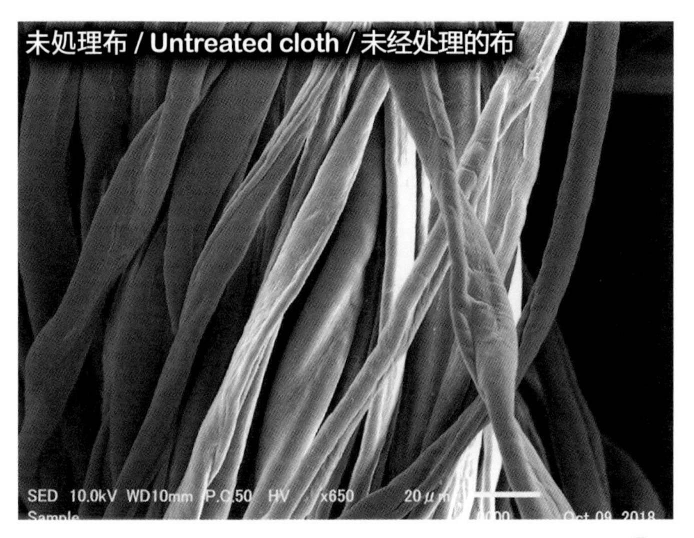

2着なので1着に換算すれば約800日間着用したことになります。

必要に応じて常温の水道水、蛍光増白剤・漂白剤不使用の洗濯洗剤を使用し、ネットに入れて他の洗濯物と一緒に洗濯機の標準コースで洗いました*。

Because it is for two clothes, so if it is converted into one pair, it corresponds to wearing about for 800 days. According to the needs, tap water at room temperature and laundry detergent without optical brightener and bleach were used. Putting it in the washing net, I washed it together with other clothes at the standard course of the washing machine.

由于是两件，因此如果将其转换为一件，则已经穿用了约800天。我们如有必要，使用室温下的自来水，并使用不使用荧光增白剂或漂白剂的洗衣粉，而是将其放入网内，而是用洗衣机的标准过程和其他衣物一起洗涤。

長期着用実験の結果 / Result of long-term wearing experiment / 长期穿着实验的结果

(1) SEM画像による観察（上の2枚の画像）/ Observation by SEM image (the two images above) / 通过SEM图像的观察（上面的两个图像）

㈱アート、伊藤社長の感想は、「こんなにべっとりと残っているとは思わなかった」。このコーティングがシルクかキトサンか、あるいはナノ銀なのか判別するために、以下の試験をしました。

President Ito's impression was, "I did not think so much coating was left". To determine if this left coating is silk, chitosan, or nano-silver, the following tests were done.

Art Co., Ltd. 伊藤社长的印象是："我没有想到还有那么多"。为了确定那该涂层是丝，壳聚糖还是纳米银，进行了以下测试。

(2) 紫外線遮蔽率試験（測定波 280~400mm）UV shielding rate test (measuring wavelength 280 to 400 mm) / 紫外线屏蔽率测试（测量波长 280 至 400 mm）

- 4年4か月着用後の加工布 / Processed cloth after wearing for 4 years and 4 months / 穿着4年4个月后经SP处理的布 ··· 92.58%
- 未処理布 / Untreated cloth / 未经处理的布 ··· 81.66%
- 上記未処理布へのキトサン加工 / Chitosan processing to the untreated cloth mentioned above / 在上述未经处理的布上进行壳聚糖加工 ··· 85.93%
- 上記未処理布へのナノ銀加工 / Nano silver processing to the untreated cloth mentioned above / 在上述未经处理的布上进行纳米银加工 ··· 83.76%

(3) 消臭試験（アンモニア）/ Deodorant test (ammonia) / 除臭测试（氨）

- 4年4か月着用後の加工布 / Processed cloth after wearing for 4 years and 4 months / 穿着4年4个月后经SP处理的布 ··· 89%
- 未処理布 / Untreated cloth / 未经处理的布 ··· 35%
- 上記未処理布へのキトサン加工 / Chitosan processing to the untreated cloth mentioned above / 在上述未经处理的布上进行壳聚糖加工 ··· 40%
- 上記未処理布へのナノ銀加工 / Nano silver processing to the untreated cloth mentioned above / 在上述未经处理的布上进行纳米银加工 ··· 31%

過去のデータによれば、シルクはアンモニア消臭効果が高く、SP加工をした綿は99%、ナイロンで80%の効果があるとの事で、上記の結果からSP加工が約90%残っていると思われます。

According to the past data, silk has high deodorant effect to ammonia, SP processed cotton has 99% of that effect and nylon has 80% of that effect. Fromm the above result, about 90% of SP processing is supposed to have remained.

根据过去的数据，据说丝绸具有很高的氨除臭效果，它具有99%的SP加工棉和80%的尼龙，从上述结果来看，我认为仍有90%的SP加工的残留。

＊他の洗濯物と一緒に：上着とバルキーセーター以外の衣服やシーツは洗濯袋に入れず、普通コースで洗い、耐久テストをしています。

＊Together with other clothes: I put my jacket and bulky sweater in a laundry bag, wash it with other laundry on a regular course, and do a durability test.

＊与其他洗衣一起：我将我的夹克和笨重的毛衣放在洗衣袋中，与其他衣物一起按常规洗涤，并进行耐用性测试。

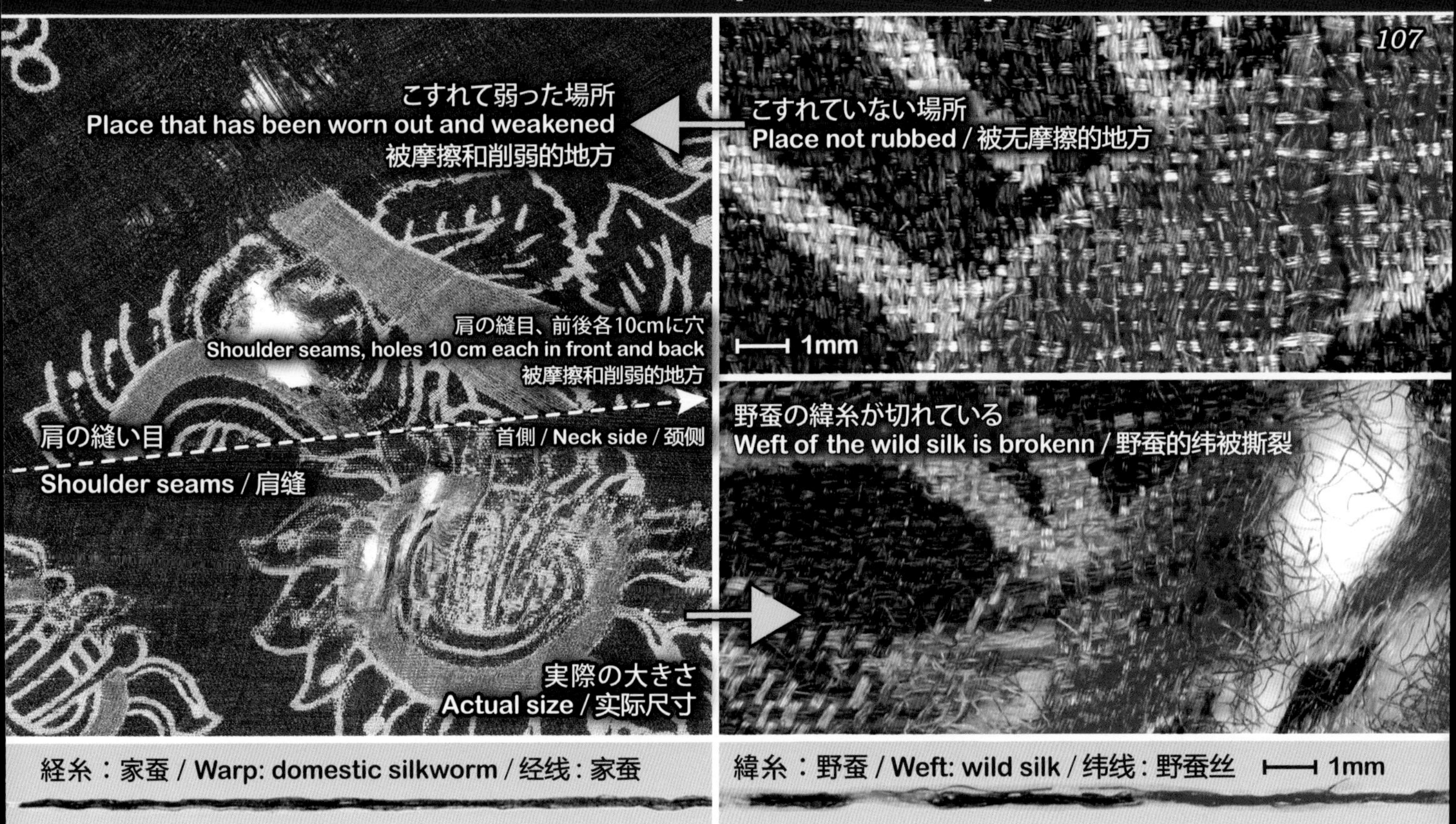

3年間着用した薄い絹製上着の肩の穴

A hole at the shoulder of a thin silk jacket worn for 3 years / 穿着3年的薄丝绸夹克的肩膀上一个洞

スカーフのように薄いプリントの上着は、インドのシルクです（上の写真）。かさばらないので、冷房で冷えたり寒風が吹く時に着るために、取材カバンに入れて持ち歩いていました。

緻密に織られた布はシワが目立たず汚れにくく、水で洗えばすぐ乾き、気になるシワはスチームアイロンですぐとれます。生地が薄いので重いカバンを、肩にかつぐのは心配でしたが、だんだん気にしなくなりました。

３年くらい経ったある日、「肩の縫い目の周りがちょっと薄くなったかなあ」と思いましたが、そのまま出張し、帰宅したら穴があいていました。「十分着た」と思いましたが、ほかの場所がちっとも古びていないので、貼りポケットをほどいて、弱った部分を修理しました。

シルクは、織り方などを選べば案外丈夫に着られます。私は、長く楽しんで着られるように手入れをしながら着用実験をしています。

Print jacket as thin as a scarf is Indian silk (pictured above). Since it is not bulky, I used to carry it in a document bag to wear when I felt cold in air conditioner or when cold wind blew.

In case of finely woven cloth, wrinkle is not easily noticed and stain-resistant. If you wash it with water, it will dry quickly, and wrinkles that you care about can be easily removed with a steam iron. I was worried about carrying a heavy bag over my shoulder because the cloth was thin, but I gradually stopped worrying about it.

Approximately three years later, one day I wondered if the seams around my shoulders might be a little thinner, but when I came home on a business trip without taking any action, I found a hole. I thought that I wore it enough, but since the other places weren't at all old, I unfastened the pasting pocket and repaired the weakened part.

Silk can be worn unexpectedly durable if you choose how to weave. I am experimenting wear while maintaining it, so that it can be worn for a long time with pleasure.

像围巾一样薄的印花夹克是印度丝绸（如上图）。由于它不笨重，因此我随身携带其放在公文包中，以便在空调中冷或吹冷风时佩戴。

精细编织的织物无皱纹和耐污渍，如果用水洗，它会很快变干，且关心皱纹可以立即用蒸汽熨斗去除。因为布很薄，我原来还担心会象扛肩膀上重的公文包那样，但我用过后，就渐渐也不再在意了。

大约三年后，有一天我想到：【肩缝周围的区域变薄了】，但是当我出差回家时，却已开了有一个洞。我以为我穿已穿得很之了，但是用由于其他地方都还没变旧，我剪开贴袋，修复了薄弱的部分。

如果选择编织方法，丝绸可能会出乎意料地耐用。我正在穿着实验，在维护的同时穿戴它以可长时间享受。

*取材カバン：軽い時でも2kg。数冊の著書を入れると、5kgくらいになります。本来、薄い上着の肩に担ぐべきではありません。

*Document bag used at coverage:
2kg even when light, If several books are in it, weight will be around 5 kg. Originally, it should not be carried over the shoulder of a thin jacket.

*公文包：即使是轻的也要2公斤，如果我放我写的几本书，则大约为5公斤。本来，它不应扛在薄夹克的肩膀上。

ウール、コットン、家蚕、野蚕、対寒耐熱温度比べ
Wool, cotton, domesticated silkworm silk and wild silkmoth silk; comparison of cold and heat resistant temperature
羊毛，棉花，家蚕和野蚕线，耐寒耐热温度比较
実験の結果、家蚕糸よりも野蚕糸の方が熱湯の熱さ、凍ったボトルの冷たさを遮断しました。右ページ参照。
As a result of the experiment, the wild silkmoth thread cut off the heat of the boiling water and the coldness of the frozen bottle more than the silk thread of the domesticated silkworm. See the page on the right.
实验的结果是，野蚕丝线比起家蚕的丝线更能切断沸水的热量和冷冻瓶的寒冷。请参见右页。
各素材で同じ重さのボトルカバーを編んでみた
Bottle covers were knitted with various materials of the same
我试图用相同重量的每种材料針织瓶套
C. 家蚕（きびそアートヤーン）
Domesticated silkworm thread (Kibiso art yarn) / 家蚕（Kibiso 真丝艺术纱）
A. ウール / Wool / 羊毛
B. 綿 / Cotton / 棉花
D. 野蚕（タサールサン：ナーシ *）
Wild silkmoth thread (Tasar silkmoth: thread of solid stem*)
野蚕线（塔萨尔蚕：nasi线*）
各糸の重さは 69g、室温 24℃ / Each thread weighs 69g, room temperature 24℃ / 每根线的重量为 69g，室温为 24℃

①3時間ごとに計測した凍ったボトルの表面とボトルカバーの表面温度
Measured temperature every three hours surface the bottle and at the bottle cover surface of the frozen water bottle
每3小时测量一次冷冻瓶子表面和瓶套的表面温度

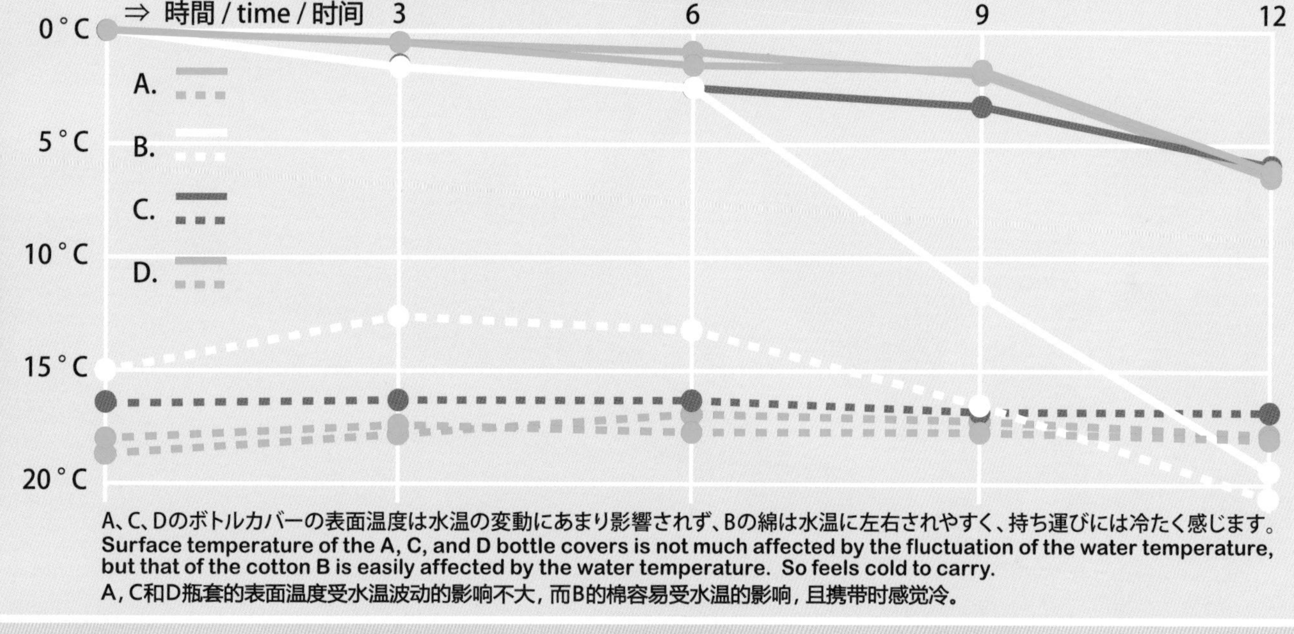

A、C、Dのボトルカバーの表面温度は水温の変動にあまり影響されず、Bの綿は水温に左右されやすく、持ち運びには冷たく感じます。
Surface temperature of the A, C, and D bottle covers is not much affected by the fluctuation of the water temperature, but that of the cotton B is easily affected by the water temperature. So feels cold to carry.
A, C和D瓶套的表面温度受水温波动的影响不大，而B的棉容易受水温的影响，且携带时感觉冷。

②3時間ごとに計測したボトル内の湯温とボトルカバーの表面温度
Hot water temperature in the bottle and surface temperature of the bottle cover measured every 3 hours
每3小时测量瓶中的热水温度和瓶套的表面温度

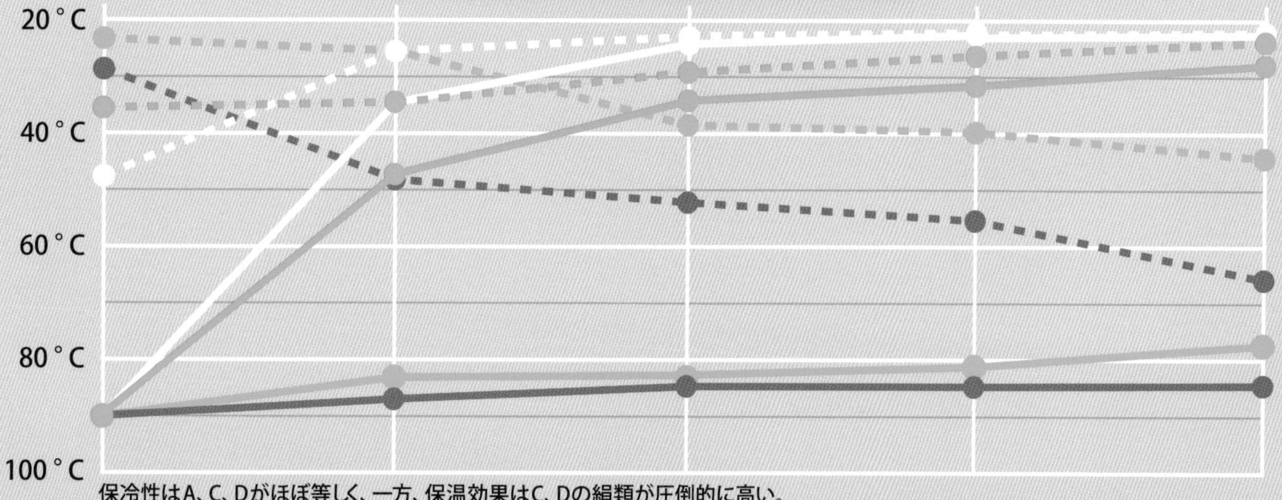

保冷性はA、C、Dがほぼ等しく、一方、保温効果はC、Dの絹類が圧倒的に高い。
Cold insulation properties of A, C, and D is almost the same level, while the heat-retaining property of the silks of C and D is overwhelmingly higher.
A, C和D的保冷性能几乎相同，而C和D的丝线的保热效果却非常高。

*ナーシ：『絹大好き２まゆの秘密』P.70 参照。
糸、アトリエ トレビ P.134。

*Nasi: See " Love silk 2 Miracle of Cocoon" P.70.
Thread:Atelier Trevi P.134.

*Nasi：请参见"爱丝2茧的奇迹"(第70页)
线程：Atelier特雷维P.134。

家蚕の日常着 / Daily wear of silkworm silk / 家蚕丝绸的日常穿着

例：着物のリメーク / Example: Kimono remake / 例如: 和服再用

縮緬 /crepe / 绉绸 *

何度も染め替えた着物の生地は弱いので、肌着などにはせず、カーテンや寝具などにします。
肌着やシーツ袋物など、こすれやすいものは縮緬で作ると長く楽しめます。
Kimono fabrics that have been re-dyed many times are weak, so do not use them for underwear, but use them for curtains or beddings.
Items that are easily rubbed, such as underwears and sheet, can be enjoyed for a long time by making them with crepe.
已多次染色的和服织物较弱，因此请勿将其用作内衣，而应将其用作窗帘或床上用品。
制内衣和床单时可使用绉绸来制作，使用時很爽滑，可长时间享受容易摩擦的物品。

縮緬 /crepe / 绉绸

縮緬 /crepe / 绉绸

縮緬 /crepe / 绉绸

＊縮緬：布幅を縮めてから服を作ると、洗濯してもしわが気にならず心地よく使えます。

＊Crepe: If you make clothes after shrinking the cloth width, you can comfortably use it without worrying about wrinkles even if you wash them.

＊绉绸：在缩小布料宽度后制作衣服，即使清洗后也可以舒适地使用它而不必担心皱纹。

シルクの手作り肌着 / Handmade silk underwear / 手工真丝内衣

肌着で締め付けられることで体調が著しく低下するときがあります。

苦労して回答が得られず、試行錯誤していたころ、絹について深く知る生活が始まりました。それから８年、絹を素肌に着ることでストレスのない生活ができることを知り、編んだり縫ったりして、肌着から外出着までを作る生活をしています。結果的に、絹の機能性をとことん追求したり、日常着の耐久テストをしたりすることになり、この本の執筆につながりました。

上の写真は、身体を締め付けないことをテーマに企画された APSARA(P.134) の肌着類です。メーカーでは、カンボジアの手織りシルクを使い、日本で作っています。

同社にはキトサンSP加工＊をしたシルクもあるので、ますます身体に優しい製品が作られることでしょう。

In case body is tightened with underwear, physical condition might deteriorate significantly. At the time when I was struggling to get an answer and doing trial and error, I learnt deeply about silk. Eight years since then, I have learned that wearing silk on bare skin will lead to a stress-free life, and I am knitting and sewing silk to make clothes from underwear to outdoor clothes. As a result, I pursued the functionality of silk thoroughly and tested the durability of everyday clothes, which led to the writing of this book.

Above picture is the underwear of APSARA (P.134), which was designed with the theme of not tightening the body. Using Cambodian hand-woven silk, they are made one by one in Japan.

Company also has Cambodia silk that has been processed with Chitosan SP*, so it will be possible to make products that are more physically friendly.

如何的身体被内衣紧束着，或许会大大地降低身体的舒适感覚。

当我在努力寻找答案并不得不反复试验时，我开始加深对丝绸的了解。八年来，我了解到在裸露的皮肤上穿丝绸可以带来无压力的生活，并且我正在编织和缝制从内衣到郊游的衣服。结果，我彻底地追求了丝绸的功能，并测试了日常衣服的耐用性，这从而编写了本书。

上图是 APSARA(P. 134) 的内衣品，其设计主题是不紧绷身体。日本制造，使用柬埔寨手工编织的丝绸在。

该公司还拥有用壳聚糖 SP ＊处理过的柬埔寨丝绸，因此有可生产出对身体更加有益的产品。

＊キトサンSP加工：加工 / アート㈱ P.138

*Chitosan SP processing: Processing / Art Co.,Ltd. P.138

＊壳聚糖SP加工：加工 / Art Co.,Ltd. 第138页。

画像提供 ①～⑦（絹 100%）：㈱シルクマルベリー (P.134)
Image provided ① ～ ⑦ (100% silk): Silk Mulberry Co., Ltd. (P.134)
提供的图片①～⑦(100%真丝)：Silk Mulberry(桑蚕丝)株式会社（第 134 页）

② 睡眠時用 / Used during sleep
睡眠期间使用

③ 紫外線対策用
For UV protection / 防紫外线

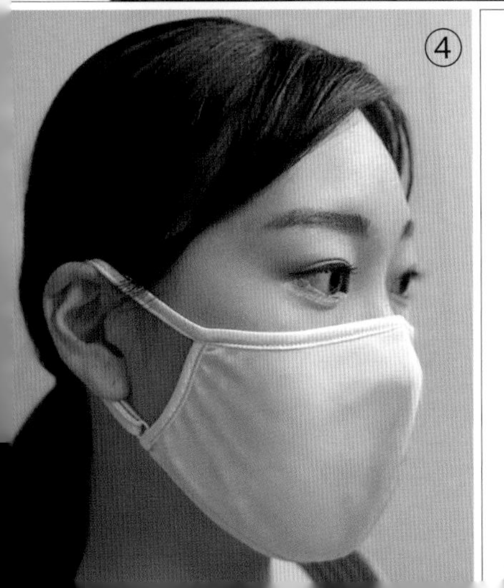

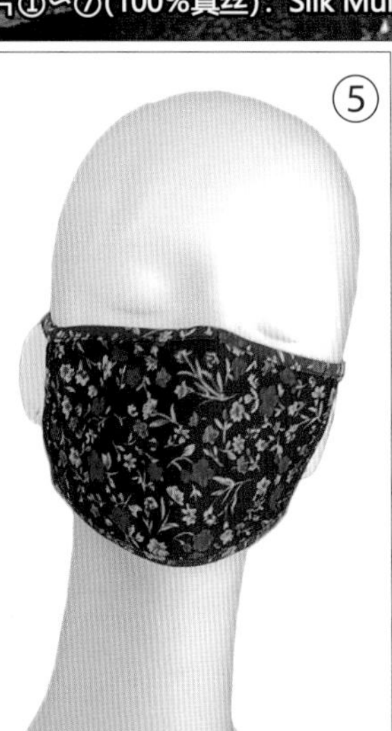

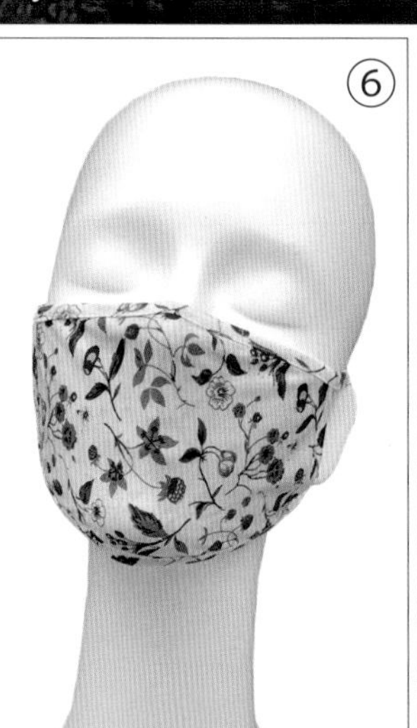

野蚕、タッサーシルクのマスク
Wild silkworm silk, tassar silk mask
野蚕丝，塔萨尔丝绸(Tassar silk)口罩

画像提供：アトリエ トレビ (P.134)。絹100%製、フィルター付き。
Image provided: Atelier Trevi (P. 134). Made of 100% silk, with filter.
提供的图片：Atelier 特雷维(P.134)。带有过滤器，100%丝绸制成。 ⑦

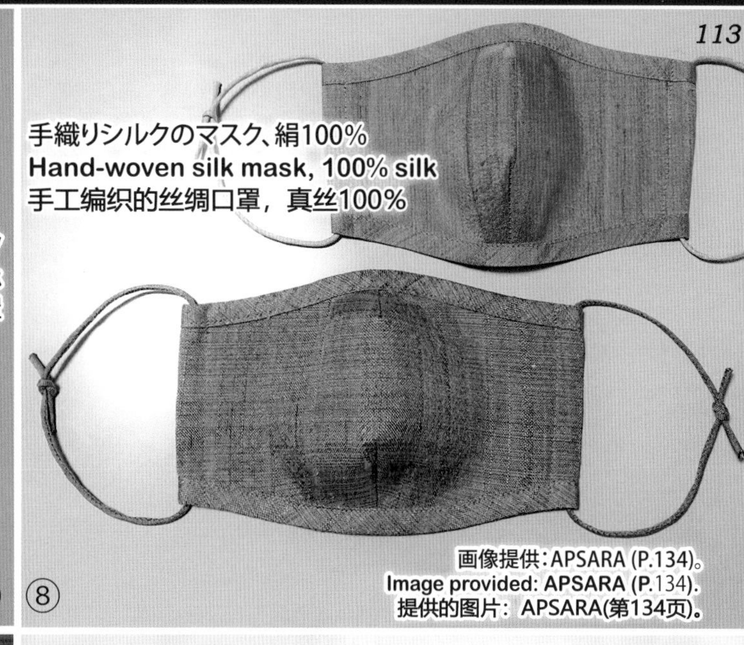
手織りシルクのマスク、絹100%
Hand-woven silk mask, 100% silk
手工编织的丝绸口罩，真丝100%

⑧ 画像提供：APSARA (P.134)。
Image provided: APSARA (P.134).
提供的图片：APSARA(第134页)。

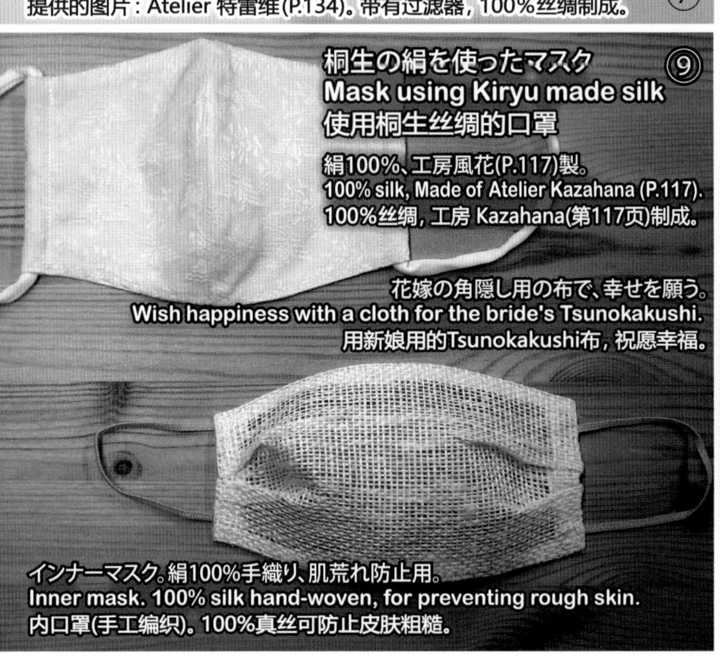
桐生の絹を使ったマスク ⑨
Mask using Kiryu made silk
使用桐生丝绸的口罩

絹100%、工房風花(P.117)製。
100% silk, Made of Atelier Kazahana (P.117).
100%丝绸，工房 Kazahana(第117页)制成。

花嫁の角隠し用の布で、幸せを願う。
Wish happiness with a cloth for the bride's Tsunokakushi.
用新娘用的Tsunokakushi布，祝愿幸福。

インナーマスク。絹100%手織り、肌荒れ防止用。
Inner mask. 100% silk hand-woven, for preventing rough skin.
内口罩(手工编织)。100%真丝可防止皮肤粗糙。

⑩ シルク裏付き、ポリエステル＆綿マスク
Silk lined, polyester & cotton mask
真丝衬里，聚酯和棉质口罩

画像提供：下山縫製㈲ (P.138)。100%シルク、キトサンSP加工のマスクも製造。
Image provided: Shimoyama Sewing Ltd. (P.138). Also manufacture 100% silk and chitosan SP processed masks. / 提供的图片：下山縫製有限公司(第138页)。比外，还生产100%真丝和壳聚糖SP处理的口罩。

日常的に使いやすい機能性 / Functionality convenient to daily activity / 易于日常生活的使用的功能

シルクマスクは、汗を吸収して放散します。このようなシルクの特性、防臭・放湿・静菌・防紫外線は、水で洗って*も消えません。

布地はウイルスや菌が透過しやすいといわれ、菌を透過させにくいといわれる不織布マスクは、マスクの中に汗をかきます。

私は不織布マスクを使用する場合、シルクの端布で作ったフィルター** を挟んでいます。

Silk mask absorbs and dissipates sweat. Characteristics of silk, such as deodorant, moisture release, bacteriostasis, and UV protection, do not disappear even when washed with water *.

Fabrics are said to be easily permeable to viruses and bacteria, and non-woven masks, which are said to be difficult to penetrate bacteria, sweat inside the mask. When I use a non-woven fabric mask, I sandwich a filter ** made of silk scrap of cloth.

丝绸口罩吸收和消散汗水。丝绸的特性，例如除臭，释放水分，抑菌和防紫外线，即使用水洗涤也不会消失*。

据说织物很容易被病毒和真菌渗透，细菌难以渗透的无纺布口罩会在面罩内流汗。

当我使用无纺布口罩时，我会放上由丝碎片布制成的过滤器**。

* 水洗い（30℃以下）何回でも使用可。
** 端切れを重ね、周囲を縫いますが、端切れのままでも案外、洗濯に耐えます。

* Can be washed with water (30 ℃ or below) as many times as you like.
** Circumference of scrap of cloth is sewed, but without sewing, it withstands washing unexpectedly.

*可以根据需要用水(30°C或以下)洗涤多次。
**缝住一块丝绸的边缘，但即使不缝制也可承受洗涤。

持続可能な開発目標 (SDGs) マスクプロジェクト
Sustainable Development Goals (SDGs) Mask Project / 可持续发展目标(SDGs)口罩项目

東京農業大学デザイン農学科では実学的観点から、学生自身が群馬県富岡市(大学と地域連携協定)の企業とコラボレーションをし、バイオベースの「人にも環境にも優しいおしゃれなマスク」を完成しました。

現在、本年末の販売を目途に商品PR　方法などに取り組んでいます(長島教授談、2020年10月)。

At the Department of Design Agriculture, Tokyo University of Agriculture, we collaborated with a company in Tomioka City, Gunma Prefecture (a regional cooperation agreement with the university), and from a practical point of view, the students themselves took the initiative to make a bio-based "fashionable and environmentally friendly mask" was produced.

Currently, we are working on product PR methods, etc. with the aim of sales start at the end of this year (Interview with Professor Nagashima, October 2020).

在东京农业大学设计农业系，我们与群马县富冈市的一家公司（与该大学的区域合作协议）进行了合作，从实践的角度出发，学生们自己主动创建了完成了基于"生物的时尚环保的口罩" 被生产了。

目前，我们正在研究产品公关方法等，以期在今年年底销售(长岛教授采访，2020年10月)。

マスク用フィルターを編んでみた / Knitted a filter for a mask / 我试过为口罩针织编过滤器

シルクに出会った2013年ころ、空調を使用して就寝するときにはマスクが必要で、耳掛けゴムではない立体成型の不織布マスクを使っていました(写真左下)。

肌触りが気になって、シルクでフィルターを編みましたが、毛羽が鼻に入るので困りました。そこで生糸のスムースニットで裏をつけましたが、やはり毛羽が気になりました。洗濯を何回か繰り返したところ、右下のKibiso糸製は毛羽がおさまり、使用可能になりました。その後、就寝時にはフェイスマスク状を使っています。

Around 2013 when I met Silk, I needed a mask when I slept under air conditioning, and I used a three-dimensional non-woven fabric mask that was not ear hanging with rubber (lower left photo).

I was worried about the texture of the non-woven fabric, so I knitted a filter with silk, but I had a trouble because the fluff got into my nose. So I backed it with a smooth knit of raw silk, but I was still worried about fluff. After repeating wash several times, fluff matter was settled in case of the filter made from Kibiso yarn of the lower right and it became ready for use.

After that, I have used a face mask type during sleep.

在我遇到丝绸时的2013年左右，我开空调睡觉时需要戴口罩，我使用了耳钩不是橡胶的，三维成型无纺布口罩（照片的左下）。

我担心无纺布的质感，所以用丝绸针织了一个过滤嘴，但由于绒毛进入我的鼻子，而给我带来麻烦。因此，我用光滑的生丝针织编物作为后盾，但我仍然担心绒毛。重复清洗几次后，右下角的Kibiso纱线因能沉降绒毛并现在可以使用它。之后，我在睡使用面罩类型。

ナーシ糸
Tasar nasi yarn / 塔萨尔nasi纱

2013 年に編み、フィルターの毛羽でむせました。
I knitted in 2013 and I choked on the fluff of the filter.
我被2013年织的过滤网的绒毛也呛过。

キビソ糸 / Kibiso yarn / Kibiso线纱

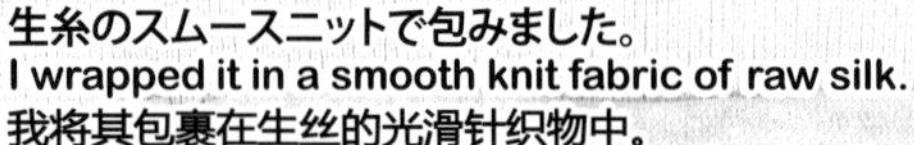

生糸のスムースニットで包みました。
I wrapped it in a smooth knit fabric of raw silk.
我将其包裹在生丝的光滑针织物中。

健康シルクの日常使用への展望 / Prospects for daily use of healthy silk / 健康丝绸的日常使用前景

半世紀前には、洋装生地店があちこちにあり、シルク100％の洋服生地や裏地が並んでいました。そして、シルクウールに化学繊維が混紡された生地はありましたが、シルクとポリエステルの混紡は記憶にありません。当時はポリエステル繊維の研究が花開き、着物や鮮やかなプリントのドレスが専門店に並び、ウオッシャブルな着物やドレスとして耳目が集まりました。それからどんどんシルクの洋装生地や裏地が消えてゆきました。

21 世紀の養蚕業を支えるためには、日常着でシルクの健康利用をひろげる必要があります。欧米では、SDGsの観点から、ポリエステルは使わないという風潮が広がっています。しかし、シルクを日常着とする場合は、ポリエステルの強靭性も必要ではないかと思います。たとえば、肌にあたる側をシルクにしたストレッチパンツやスパッツ。日々シルクの特性を得ながら、こすれに弱いなどのシルクの弱点をポリエステルがカバーします。

Half a century ago, there were many western-style fabric store, which displayed Silk clothing fabrics and linings for western style clothes. And although there was a fabric in which chemical fibers were blended into silk wool, I don't remember the blending of silk and polyester. At that time, research on polyester fibers flourished, and kimonos and brightly printed dresses were lined up in specialty stores, attracting attention as washable kimonos and dresses. After that, the silk cloth and silk lining disappeared more and more.

In order to support the sericulture industry in the 21st century, it is necessary to expand the healthy use of silk in everyday wear. In Europe and the United States, there is a growing tendency not to use polyester from the perspective of SDGs. However, the toughness equivalent to polyester is required to use silk for everyday wear. For example, stretch pants and spats with silk on the side that touches the skin. By using polyester, it is possible to cover the weak points of silk such as being easily rubbed and bring out the characteristics of silk.

半个世纪前，有许多西式面料商店，两旁都是丝绸服装面料和衬里。还有一种织物，其中化学纤维混纺到了羊毛中，但是我不记得丝绸和聚酯的混纺。当时，对聚酯纤维的研究蓬勃发展，和服和鲜艳印花的连衣裙在专卖店排起了长队，作为可洗的和服和连衣裙引起了人们的关注。此后，真丝西式面料和真丝衬里越来越不见了。

为了支持 21 世纪的养蚕业，有必要扩大在日常穿着中丝绸的健康利用。在欧洲和美国，从可持续发展目标的角度来看，越来越多的趋势是不使用聚酯。然而，在日常穿着的丝绸中，还是很需要有聚酯的韧性。例如，弹力的裤子和打底裤的接触皮肤的一侧使用丝绸。通过使用聚酯，在每天获得丝绸特性的同时，在外面通过使用聚酯，可以覆盖丝绸的弱点，例如不耐磨 ，并展现出丝绸的特性。

シルクの特性で特許取得、研磨剤フリーの練り歯磨き
Abrasive-free toothpaste, patented for silk properties / 无磨蚀性牙膏，拥有丝绸特性的专利

歯表を削ることなく、ホワイトニングや電動歯ブラシを使用したい人々への朗報です＊。

東京農業大学長島研究室で、「エリシルクが歯の汚れを吸着し、エナメル質を傷つけることなく歯を自然な白さにする」ことを実証した結果、オーガニクス オーラルケア製品が誕生しました（2019年11月販売開始､写真下）。発泡剤、増粘剤フリー。

子どもから大人、妊産婦にも安心な、メイド オブ オーガニクス ホワイトニング トゥースペースト シルクパウダー。

歯磨き後も食事の味に影響を与えにくく、天然由来成分 100%。

製造販売元。㈱たかくら新産業 (P.138)。
https://www.madeoforganics.com/

Good news for people who want to use whitening and electric toothbrushes without scraping the tooth surface.

As a result of demonstrating that "Erisilk absorbs stains on teeth and makes teeth natural white without damaging enamel" at the Nagashima Laboratory of Tokyo University of Agriculture, an organic oral care product was born (2019. Started selling in November, bottom photo).

Free of foaming agents and thickeners. Whitening toothpaste of "Made of Organics" containing silk powder, which is safe for children, adults and pregnant women. 100% naturally derived ingredients that do not easily affect the taste of meals even after brushing teeth.

Manufactured and sold by TAKAKURA NEW INDUSTRIES INC. (P.138).
https://www.madeoforganics.com/

对于想使用美白和电动牙刷而不刮擦牙齿表面的人来说是个好消息＊。

由于在东京农业大学长岛实验室证明“ Erisilk 吸收牙齿上的污渍并使牙齿自然变白而不会损害牙釉质 ”，因此诞生了有机口腔护理产品（2019年。11月开始销售，左图）。

不含发泡剂和增稠剂。含有蚕丝的“Made of Organics”的美白牙膏，对儿童，成人和孕妇均安全。100%天然来源的成分，即使刷牙后也不会轻易影响餐食的味道。

高仓新产业株式会社制造和销售（第 138 页）。
https://www.madeoforganics.com/

＊同社ホームページによれば、エリシルクの配合により歯垢やステインなどの汚れを吸着して浮かせ、ブラッシングにより取り除きやすくし、歯にやさしく、自然な白さに導くとのこと。

＊According to the company's website, the combination of Erisilk adsorbs and floats stains such as plaque and stains, and brushing makes it easier to remove, leading to tooth-friendly and natural whiteness.

＊根据该公司的网站，Erisilk的组合可以吸附和漂浮牙菌斑和污渍等污渍，刷牙更容易去除，从而带来对牙齿友好和自然的白度。

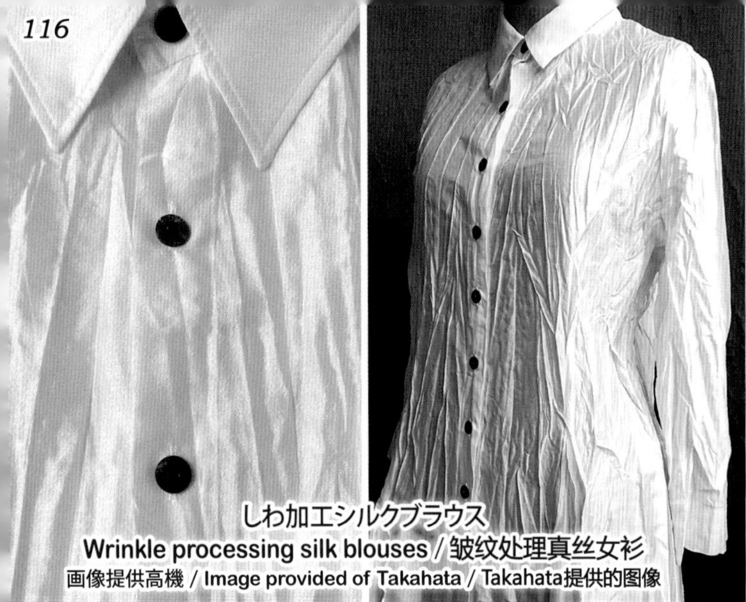

しわ加工シルクブラウス
Wrinkle processing silk blouses / 皱纹处理真丝女衫
画像提供高機 / Image provided of Takahata / Takahata提供的图像

縮緬よりも細かい織地の凹凸は超強撚糸によって織られ、快眠に誘う。
Asperity of the fabric woven with very strongly twisted yarn, is finer than that of crepe, and induces sound sleep.
比绉绸更细的机织物不均匀性是由超强捻纱织成的，对睡眠有利好觉。

左上写真はシルク70%、ポリエステル30%の紗（しゃ）織*です。群馬県桐生市の下山縫製の高機事業部製。左下のキビソのストールを織った手織り機の名称、高機が事業部名。山車幕などの伝統工芸品も制作しています。

桐生市ではシルクの日常使用のため、伝統の桐生織りを日用品にする試みがあり、右上写真は八丁撚糸**を使ったお召し織りのシーツです。高級な着物地の広幅は70cm、縫い合わせ目があります。

シルクを大量に使用する日用品に、毛布があります。右下は柞蚕の毛布、カシミア毛布くらいの価格ですが、エアコン下で夏冬使えます(アトリエ トレビ調べ)。絹の寝具が日用品として普及すると、養蚕業の未来が見えてくるでしょう。

Upper left photo is a "Sha(Leno)" weave* with 70% silk and 30% polyester. Made by the Takahata Business sector of at Shimoyama Sewing(P.135) in Kiryu City, Gunma Prefecture. Takahata is name of the hand-woven machine that weaves the kibiso shawl (lower left), and the name of the Business sector. Traditional crafts such as "Float curtains" are also made. In Kiryu City, there is an attempt to use traditional Kiryu weave as a daily necessities for the promotion of silk daily use, and the upper right photo is a bed sheet of Omeshi woven using Haccho twisted yarn **. Width of the high-class kimono fabric is 70 cm, and there are seams.

Blanket are one of daily necessities that use a large amount of silk. Lower right is the Tussah silk blanket whose price is almost equivalent to that of cashmere blanket, but it can be used both in summer and winter under air conditioning (according to Atelier Trevi). When silk bedclothes will become popular as daily necessities, the future of the sericulture industry will be prospective.

左上方的照片是用70%的丝绸和30%的聚酯制成的【沙罗织物】*。由群马县桐生市Shimoyama缝纫(第135页)的Takahata事業部制造。Takahata是编织kibiso披肩（左下）的手动织机的名称，也是部门的名称。他们还生产传统工艺品，例如【彩车窗帘】。在桐生市，有一种尝试将传统的桐生织法用作日常使用的丝绸，而右上方的照片是使用Haccho加捻纱**织成的Omeshi(熟丝绉绸)的床单。高级和服面料的宽度为70厘米，并且有接缝。

使用大量丝绸生产的日常产品中，有毯子。右下方是柞蚕丝毯子，价格大和羊绒的价格大致一样，可在夏季至冬季在空调下使用(根据Atelier特雷维)。当丝绸床上用品成为日常必需品时，我认为我们可以看到蚕业的未来。

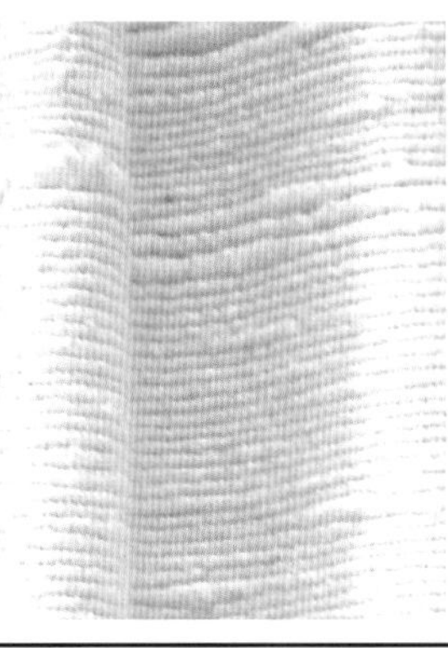

経糸：生糸。緯糸：キビソ、手染め。
Warp: raw silk, weft: Kibiso, hand-dyed.
经线：生丝。纬线：Kibiso，手工染色。

幅 170cm× 長さ 210cm
Width 170 cm x length 210 cm / 宽 170 厘米 x 长 210 厘米

毛羽部：基布：緑布：縫い糸まですべて高品質の柞蚕シルク100%。
Fluff: Base cloth: Edge cloth: 100% high quality Tussah silk, including sewing thread.
绒毛：基布：边缘布：100%优质柞蚕丝，包括缝纫线。

*紗織：伝統の桐生織りの「もじり織」の一種。
**八丁撚糸：桐生では 1990 年代に作られなくなり、八丁ヤーン㈱により復活した強撚糸。

*Leno weave: A type of "Mojiri weave", traditional Kiryu weave.
**Hatcho Twisted Yarn: A strong twisted yarn that was no longer made in Kiryu in the 1990s and was revived by Hatcho Yarn Co., Ltd.

*沙罗织物：一种传统的桐生织法"Mojiri织法"。
**Hatcho加捻纱：一种强捻纱，1990年代在Kiryu不再生产，并由HatchoYarn Co., Ltd.恢复。

身近なシルクの展示 / Exhibition of familiar silk / 熟悉的丝绸展示

ワイルドシルクミュージアム / WILD SILK MUSEUM / WILD SILK(野生丝绸)博物馆

シルクがもっと身近に感じられる場所として開設。家蚕と野蚕のシルク製品を、からだに優しく健康的な繊維として展示販売もしています。

〒135-0023 東京都江東区平野 1-5-5-101
http://www.wildsilk.jp

Established as a place where silk can be felt more closely. We display and sell silk products of domesticated silkworms and wild silkmoths as healthy fibers that are kind to the body.

1F 1-5-5 Hirano, Koto-ku, Tokyo 135-0023, Japan http://www.wildsilk.jp

开设为可以更紧密地感觉到丝绸的地方。
展示和销售家蚕和野蚕的蚕丝产品，作为对人体友善的健康纤维。

1F 1-5-5 Hirano, Koto-ku, Tokyo 135-0023, 日本 http://www.wildsilk.jp

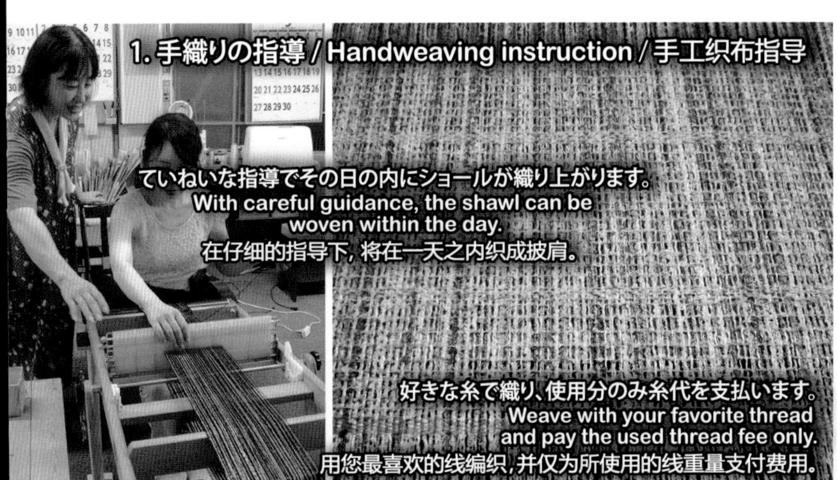

1. 手織りの指導 / Handweaving instruction / 手工织布指导

ていねいな指導でその日の内にショールが織り上がります。
With careful guidance, the shawl can be woven within the day.
在仔细的指导下，将在一天之内织成披肩。

好きな糸で織り、使用分のみ糸代を支払います。
Weave with your favorite thread and pay the used thread fee only.
用您最喜欢的线编织，并仅为所使用的线重量支付费用。

2. 繭毛羽からの糸づくり
Making threads from floss of cocoon
用茧衣的制作纱

実際の大きさ
Actual size / 实际尺寸

ガラ紡
Gara spinning machine
Gara 纺机

工房 風花 (絹遊塾)* / Atelier Kazahana (Silk Schools)* / 工房 Kazahana (丝绸学校)*

手染めのオリジナル糸が豊富で、こんな物を織りたい、こんな色を染めたい、こんな物を縫ってみたい、そんな気持ちを応援してくれる工房です。

一方、ガラ紡で繭毛羽から糸を紡ぐ、珍しい作業も見学できます。

There are abundant hand-dyed original threads, and it is a workshop that supports the desire to weave such things, dye such colors, and sew such things following your desire. Other hand, you can see the rare manufacturing process of spinning yarn from floss of cocoon with Gara spinning machine.

手工染色的原始线很多，这是一个支持编织此类东西，染上此类颜色并缝制此类东西的工房。另一方面，您可以观察到用Gara纺机丝到的稀有的从茧衣中纺出的纺丝。

＊〒376-0053 群馬県桐生市東久方町 1-1-55
ベーカリーレンガ付設工場内 ☎0277-32-6387
https://www.kiryu-renga.jp/kazahana

＊Within Bakery Renga attached factory 1-1-55 Higashi hisakata-cho, Kiryu, Gunma 376-0053, Japan
https://www.kiryu-renga.jp/kazahana

＊Bakery Renga laying factory 1-1-55 Higashi hisakata-cho, Kiryu, Gunma 376-0053, 日本
https://www.kiryu-renga.jp/kazahana

1964年開設。2014年8月岡谷蚕糸博物館、愛称「シルクファクトおかや」としてリニューアルオープン。館内に㈱宮坂製糸所を併設、諏訪式繰糸機、上州式繰糸機、自動繰糸機などを動態展示。

今ではあまり見ることのできない伝統的な糸取りから最先端の繰糸法までが、一年中開館時間内に見学できます。

Established in 1964. In August 2014, it was reopened as Okaya Silk Museum nicknamed "Silk Fact Okaya". Miyasaka Silk Mill Co., Ltd. is set up in the hall, and Suwa type reeling machine, Joshu type reeling machine, and automatic reeling machine, etc. in operation are exhibited. Reeling method from traditional one, which is rarely seen now, to the most advanced one, can be seen all year round within the opening hours.

成立于1964年。2014年8月，它被重新开放为绰号Okaya Silk Museum的【Silk Fact Okaya】。在展厅内设立宫阪丝绸厂有限公司，并动态展览Suwa式缫丝机，Joshu式缫丝机和自动缫丝机等。从现在很少见的传统缫丝方法到最先进的缫丝方法，可以在全年的开放时间内供参观。

青梅きもの博物館 / Ome Kimono Museum / 青梅和服博物馆*

皇室や江戸時代の貴重な衣装(着物)が収蔵されています。

21世紀の今では見ることが難しい、限られた人々に提供された技術の粋を集めた着物が間近に見られます。

It houses valuable costumes (kimono) of the Imperial family and at Edo period. Collection of Kimonos that were made up with the best of technology and were provided to a limited number of people can be seen closely. They are difficult to see at present in the 21st century.

它容纳了皇室和江户时代的贵重服装【和服】。
可以看到汇集了最先进技术的和服提供给有限数量的人，这在21世纪现在很难看到。

＊〒198-0063 東京都青梅市梅郷4-629 ＊4-629 Baigo, Ome, Tokyo 198-0063, Japan ＊4-629 Baigo, Ome, Tokyo 198-0063, 日本

あ・ア A

い・イ I

う・ウ U

え・エ E

お・オ O

か・カ Ka

き・キ Ki

く・ク Ku

け・ケ Ke

こ・コ Ko

さ・サ Sa

し・シ Si (Shi)

す・ス Su

せ・セ Se

そ・ソ So

た・タ Ta

A

B

C

D

U

V

W

Y

M

N

P

Q

R

S

T

U

W

X

後援 / Sponsors / 赞助商

皆様のおかげでこの本が出版できました / Thanks to everyone, this book could be published
多亏了大家，我们了才得以出版这本书

個人
Individual / 个人的

台湾では健康のために絹を着ます
Taiwanese wear silk for health
我在台湾为健康而穿着丝绸

黄 麗珠
Kou Reishu

台湾 台北市
Taipei County, Taiwan

ワイルドシルク専門店
Shop specializing in Wild silk / 野蚕丝绸专卖店

アトリエ トレビ / Atelier Trevi / Atelier 特雷维

素材から製品まで
MATERIAL to CLOTHING
从材料到产品

〒101-0052 東京都千代田区神田
小川町2-8 扇ビル3F

☎ 03-6904-2560 (+81-3-6904-2560)
http://www.akasaka-trevi.jp
Ougi Building 3F
2-8 Kanda Ogawamachi, Chiyoda-ku,
Tokyo 101-0052, Japan

シルク専門メーカー「主に家蚕」
Silk specialized manufacturer "Mainly mulberry silk" / 丝绸专长制造商【以家蚕为主】

大変難しいシルクの縫製は、提携工場の選ばれた縫製エキスパートが担います
Very difficult silk sewing is taken care of by selected sewing experts of a partner factory
最困难的丝绸缝制是由合作工厂的选定缝制专家进行的

私達は、シルクで心と身体がリラックスすることを目指しています
We aim to relax both body & mind with our Silk products / 我们的目标是用我们的丝绸产品放松身心

衣料品から寝具まで
From clothing to bed clothes
从服装到床上用品

㈱シルクマルベリー／ Silk Mulberry Co., Ltd.

〒270-0014　千葉県松戸市小金 217-2　☎ 047-345-4695（+81-47-345-4695）
https://www.silkmulberry.com　217-2 Kogane Matsudo, Chiba 270-1114, Japan

「カンボジア手織りシルク」取り扱いメーカー
Maker handling "Cambodia hand-woven silk" / 厂商处理【柬埔寨手织丝绸】

「水洗いできる手織りシルクの手作り衣」/ "Handmade clothes from hand-woven silk that can be washed with water"
【可以用水洗的手工编织的丝绸手工衣服】

NPO Shien Tokyo アプサラ / APSARA

〒177-0034 東京都練馬区富士見台 1-19-5-202
☎ 03-3577-5026 (+81-3-3577-5026)
https://apsara.shopselect.net
1-19-5-202 Fujimidai, Nerima-ku,
Tokyo 177-0034, Japan

ショール・アクセサリーも！
Also shawls and accessories!
还有披肩配件!

絹製手作りペルシャ絨毯（オリジナル デザイン）
Silk handmade Persian carpet (original design) / 丝绸手工波斯地毯 (原创设计)

世界で1枚 / Only one in the world / 世界上只有一件

エイワン ピルモラディ
EYWAN PIRMORADI

〒112-0013 東京都文京区音羽1-15-15 シティ音羽 1F
☎ 03-3941-4850(+81-3-3941-4850)
http://www.eywan.net/
City Otowa 1F 1-15-15 Bunkyo-ku,
Tokyo 112-0013 Japan

食べるシルク・レストランウエディング・本格イタリアンダイニング
Edible silk protein, restaurant wedding, authentic Italian dining / 食用丝蛋白，餐厅婚礼，正宗的意大利美食

カフェ＆ダイニングイースト／CAFE&DINING EAST（open 10:00~21:00）

「庄内シルク物語」期間 6月1日～30日ディナーのみ「要予約」/ A コース¥2,500. B コース¥3,500.
"Shonai Silk Story" June 1 to 30、Only dinner, "reservation required" / A Course ¥2,500. B Course ¥3,500.
【Shonai 丝绸故事】6月1日至 30 日：仅提供晚餐(需要预订) / A 套餐 2500 日元，B 套餐 3500 日元

〒997-0801 山形県鶴岡市東原町17 グランド エル・サン内
☎ 0235-24-4633（+81-235-24-4633）
https://cafe-east.net/
In Grand el Sun 17-7 Higashiharamachi, Tsuruoka,
Yamagata 997-0801 Japan

界面活性剤を始めとする、各種工業用薬剤の製造・販売
Manufacture and sale of various industrial chemicals including surfactants to be the lead

「まゆ」から出発した百年企業 / Our century from a cocoon

第一工業製薬は明治42年、1909年に京都で創業しました。お菓子のおいしさを引き出す材料から人工衛星が宇宙で動く電池の部材まで、人間生活に欠くことのできない商品を作っている会社です。百年前の日本の国家的産業は絹織物でした。その絹糸の基になるカイコのまゆをほぐしやすくする液体を発明します。当時のベンチャー企業でした。

水と油は、交わらないものの代表です。仲のよくない水と油を交わらせる物質を界面活性剤といいます。誰もが知っている石鹸が代表例です。明治の時代でありながら、国際的にシルク・リーラーと名づけて売り出していました。カイコは、この会社の創業を担う大切な役割を果たしてくれたのです。現在の当社の多くの技術の基本となっています。

産業は、第4次革命に入ったと言われます。動物の中で道具を使うのは人間だけであり、進歩させて来ました。人工知能で、道具は最終段階を迎えたと考えています。データの蓄積と計算では、人間の能力を超えました。シンギュラリティ、人工知能が人間を超える限界。機械が人間を超えるでしょうか。人間は人間であり、機械に人間性はありません。

産業発展により拡大した経済は、人間の感性や欲望の合計数です。人間があるから経済があります。快さや幸せを求める人間の原点は、健康志向です。産業進化で発生した環境汚染を解決するリサイクルの象徴は植物です。桑と共に生きるカイコには、共生のヒントがあります。快適に長生き。日本発の和医学の事業化を百年企業が取り組んでいます。

DKS started its business in 1909 in Kyoto. We are the company making a variety of products that are essential to human life, from a material to enhance tastiness of cakes to a battery component for satellites so that they can function well in the space. A century ago when silk textile manufacturing was the biggest industry of Japan, our founders invented a liquid to make unraveling of silk yarns from cocoons easier and quicker. It was a venture company at the time.

"Water and oil" is a typical combination of substances which are incompatible, which cannot get along together. However, with a help of a "surface-active agent", or surfactant, they can be mixed. The most well-known surfactant is soap. DKS launched its first surfactant product under the English trade name SILK-REELER so that it could be marketed internationally in the Meiji period. Silkworms played an important role in our foundation. SILKREELER is still at the basis of many of our present technologies.

創業時のぼり旗「シルクリーラー」
Banner-flag at the time of establishment "SILKREELER"

We are now in the fourth industrial revolution. We, the human beings, are the only animal to use tools, and we have been improving them throughout our history. By the emergence of artificial intelligence (AI), I believe that the tools we make have reached their last phase of improvement. AI has already exceeded the human ability in data accumulation and calculation. Singularity： *The limit where AI surpasses the humans.* Well, can machines really surpass humans? We have humanity and only humans can have humanity.

Economy is the aggregate of human emotions and aspirations. We seek for happiness and comfort, and the most fundamental desire of human beings is that for HEALTH. While the industrial revolutions brought about environment pollutions, plants are a symbol of recycling which would save the environment. Silkworms in mulberry trees give us hints for coexistence. *For long and comfortable living*, the century-old company is now working on the field of traditional Japanese medicine originated from Japan.

代表取締役会長兼社長　坂本 隆司 / SAKAMOTO Takashi, Chairman CEO

第一工業製薬株式会社 / DKS Co. Ltd.

〒601-8391 京都市南区吉祥院大河原町5　☎ 075-323-5911（+81-75-323-5911）FAX 075-326-7356（+81-75-326-7356）
5 Ogawara-cho, Kisshoin, Minami-ku, Kyoto 601-8391, Japan　https://www.dks-web.co.jp/

シルクタンパク質製造販売「食べるシルクタンパク質」
Silk protein production and sales; "edible silk protein" / 蚕丝蛋白的生产和销售【食用蚕丝蛋白】

子どもたちの未来に愛を込め、ゆるぎない心で未来を拓く

ドクターセラムは平成17年、2005年に創立しました。現在当社のメイン商材である「シルクフィブロイン」は、東京農業大学 長島孝行 農学博士と会話をした中で、シルクタンパク質の機能が私自身にとって非常に有用であることを知り、試行錯誤の結果、製品として完成しました。以来、日々愛飲しています。私は玉石混交の業界の中で、いつも家族が使うことを念頭に、「本物」といえる素材や商材を探求し続けることに労を惜しみません。

代表取締役　吉川 育矢

We continue to open up the future of children with love and unwavering resolve
将爱注入儿童的未来，以坚定不移的心开启未来

Doctor Serum Co., Ltd. was founded in 2005. “Silk Fibroin”, main product of our company was turned into a commercial reality through a trial and error process, since I found its function very important for my own health through the conversation with Dr. Takayuki Nagashima, Tokyo University of Agriculture. Since then, I have continued to take that product every day. I am always willing to continue my search for materials that can be said to be "real" within the industry where the mixture of wheat and chaff exists, taking the use of the family into consideration.

Doctor Serum株式会社成立于2005年。通过与东京农业大学长岛孝之博士的交谈，我发现它的功能对我自己的健康非常重要，因此我们公司的主要产品【丝素】通过反复试验将其转化为商业现实。从那时起，我每天都继续服用该产品。在善恶混杂的行业工作中，我会不遗余力地继续寻找可以说是【真品】的材料和产品，时刻牢记家人会用的。

YOSHIKAWA Ikuya, Representative Director / 代表董事 吉川育矢

セラム-シルクフィブロインは、アジアゴールデン スター アワード 2017 で、優れた品質と革新的な製品や商品であり際立って革新性が有ると認められた商品に贈られる”商品賞”を受賞しました。

Doctor Serum Co., Ltd was awarded "Goods Award" for SERUM - Silk fibroin at Asia Golden Star Award 2017. That award is given to excellent quality & innovative products with distinguished innovation.

SERUM丝素蛋白因其卓越的品质和创新的产品，被公认为杰出创新性的产品，而获得了2017年亚洲金星奖的【产品奖】。

ドクターセラム㈱／ Dr. Serum Co., Ltd.

http://www.dr-serum.com

〒150-0043 東京都渋谷区道玄坂1-22-8 4F
☎ 03-5728-8825（+81-3-5728-8825）
4F 1-22-8 Dogenzaka, Shibuya-ku, Tokyo 150-0043, Japan

「絹生活研究所」が無菌「みどり繭」で新しいシルク産業の未来を創る
Skin life lab make the future of new silk industry by a special cocoon "Midorimayu" growing in aseptic room
皮肤生命实验室通过在无菌室中生长的特殊茧【Midorimayu】使新丝绸工业的未来发展

Silk life lab.
絹生活研究所のシルク製品「みどり繭」の力で人びとの健康と美を守る

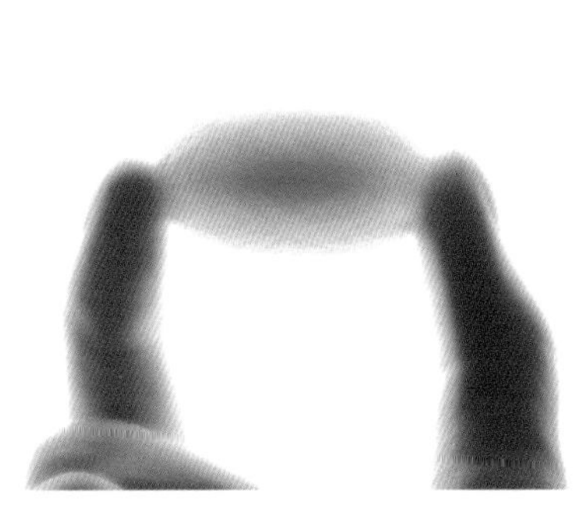

絹生活研究所は「みどり繭の力で人びとの健康と美を守る」をコンセプトに、コスメティック・ファブリック・サプリメンの3カテゴリで幅広く商品を展開するライフスタイルブランドです。

Silk product of Silk life lab protect people's health and beauty by power of "Midorimayu"
丝绸生活实验室的丝绸产品通过【Midorimayu】的力量保护人们的健康和美丽

Silk life lab's concept is to protect people's health and beauty by power of "Midorimayu". This is the life style brand developing extensive products in three categories, they are cosmetics, fabric, and supplement.

丝绸生活实验室的理念是通过【Midorimayu】的力量保护人们的健康和美丽。这是一家生活品牌，开发化妆品，织物和补品这三类产品。

シルクの中でもとりわけ希少価値の高い「みどり繭」を世界で初めて無菌養蚕工場で生産することに成功しました。みどり繭は、長年の研究によって品種改良され、餌である桑の葉の色素が抽出された繭です。美しい機能的なシルクの特性に加え、健康成分・フラボノイドを一般的な白繭の2倍以上含有した「健康になるためのシルク」いえます。絹生活研究所のすべての製品には、自社システムで生産される安心・安全なみどり繭を使用しています。

We succeeded in producing "Midorimayu" especially having a higher scarcity value than the other cocoon at aseptic sericulture factory for the first time in the world. Midorimayu is the cocoon bred in the long years of research and having pigment extracted from mulberry leaf which is silkworm's feed. In addition to beautiful and functional properties of silk, it contains flavonoid which is healthy ingredient more than twice as many as the general cocoon. It is the silk to get healthy. All products of Silk Life lab use secure and safe Midorimayu which produced in artificial aseptic sericulture factory.

我们在世界上首次成功地生产了【Midorimayu】，它的稀缺性价值比无菌养蚕桑厂中的其他茧高。Midorimayu是经过长期研究而培育的茧，其色素是从桑叶中提取的，桑叶是蚕的饲料。除了丝绸的美丽和功能特性外，它还含有类黄酮，它是健康成分，是普通茧的两倍以上。是健康的丝绸。Silk Life实验室的所有产品都使用安全，安全的Midorimayu，它是由人工无菌养蚕桑厂生产的。

Silk life lab.
絹生活研究所

■株式会社きものブレイン／ Kimono Brain Co., Ltd.
夢ファクトリー本社工場／ Dream factory and head office http://www.kimono-brain.com
〒948-0056 新潟県十日町市沢口丑 510-1 ／ 510-1 Sawakuchi ushi,Tokamachi-shi,Niigata,948-0056,japan
代表 TEL(025)752-7700 FAX(025)757-2008 Main number TEL. +81-25-752-7700 FAX.+81-25-757-2008

生活に新しい発想と選択肢を提供する企業
Companies that provide new ideas and options for life / 提供生活新思路和新选择的公司

ナチュラルとオーガニックを、デイリーユースへ / Natural and organic products for daily use / 日常使用的天然和有机的品

“かぞく”へ 安心して手わたせるもの

Things that can be safely handed to "family"/ 可以安全地交给【家庭】来使用

株式会社 たかくら新産業
TAKAKURA NEW INDUSTRIES INC.
高仓新产业 株式会社

https://takakura.co.jp

日本製ベビー用品メーカー
Baby products made in Japan / 日本制造婴儿用品制造商

「テンサン」シルク化粧品専門メーカー
"Tensan" silk, cosmetics specialized manufacturer /【天蚕】丝绸化妆品生产商

㈱ファランドール / Farandole Co., Ltd.

〒222-0033 神奈川県横浜市港北区
新横浜 2-6-3 DSM新横浜ビル3F
☎ 045-471-5888 (+81-45-471-5888)
http://www.mayuco.jp
DSM Shin-Yokohama Building 3F
2-6-3 Shinyokohama, Kohoku-ku,
Yokohama-shi, Kanagawa
222-0033 Japan

シルクタンパク質製造販売「食用」「化粧品用」「繊維加工用」
Production and sale of silk protein "edible" "for cosmetics" "for textile processing" / 蚕丝蛋白的生产和销售【食用】【化妆品用】【纤维加工用】

シルクタンパク質の原液の製造販売
シルク入り基礎化粧品・石鹸・ネイルローションの企画・製造・販売

Manufacture and sale of undiluted solution of silk protein: Planning, Manufacture and sale of basic cosmetics, soap and nail lotion containing silk / 生产和销售未稀释的丝蛋白溶液：开规划和生产和销售，含丝的基础化妆品，肥皂和指甲乳液

ネールローション
Nail Lotion / 指甲乳液

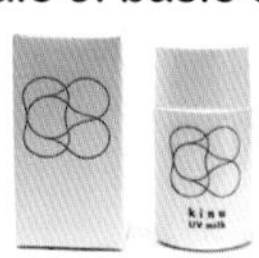

基礎化粧品
Basic cosmetics / 基础化妆品

手作り石鹸
Handmade soap / 手工皂

㈱アート / ART Co.,Ltd.

〒376-0011 群馬県桐生市相生町 2-620
☎ 0277-54-5178 (+81-277-54-5178) http://art-silk.jp
2-620 Aioi-cho, Kiryu, Gunma 376-0011, Japan

冷えとりくつ下と天然素材の店
Shop of cold prevention socks and natural materials / 防寒袜和天然材料的商店

繭結
MAYU

〒177-0034 東京都練馬区富士見台 2-1-11
☎ 03-5848-3917 (+81-3-5848-3917)
2-1-11 Fujimidai, Nerima-ku,
Tokyo 177-0034, Japan
https://twitter.com/mayu_fujimidai

不可能を可能に、今までにない「縫製店」を目指す企業
A company that makes the impossible possible and aims to become an unprecedented "sewing shop"
将不可能变为可能致力于成为前所未有的【缝纫店】的公司

各種プリーツ・シワ加工・日本古来の紗・その他
Various pleats, wrinkle processing, ancient Japanese gauze, etc.
各种褶，皱纹处理，日本古代纱織布等

下山縫製㈲ / Shimoyama sewing Ltd. / 下山缝纫㈲

〒376-0011 群馬県桐生市相生町1-228
☎ 0277-54-5281 (+81-277-54-5281)
1-228 Aioi-cho, Kiryu, Gunma
376-0011 Japan
https://shimoyamasilk.jimdofree.com

シルクの持続可能な開発目標 (SSDGs)
Silk Sustainable Development Goals (SSDGs)
丝绸的可持续开发目标 (SSDGs)

シルクは、欧米の人々にとってSDGsに反する産物とされています。 しかし、生育が早く傾斜地にも根づく桑は、カイコの餌となるとともに荒れた山岳地帯を修復する植物となる可能性があります。

果実はジャムやジュースになり、葉は副菜になり、お茶にもなる。江戸時代の末期まで桑葉は糖尿病の民間薬でした。

近年の研究によれば、食後の血糖値上昇を穏やかにする効果が糖尿病の予防につながるとして期待されています。

大木になれば、木部は美しい木目の木材になります。給餌後の枝や選定後の枝は、健康食品の原料や和紙になる可能性があります。

ウイズコロナの時代。幼稚園、小学校、老人ホーム、過疎地域での桑の栽培や養蚕活動を促進する養蚕施設など、桑や蚕が生み出すさまざまな自然の恵みは、人間らしい新しい生き方を生み出す可能性を秘めています。

シルクの長い歴史は、持続可能であったからこそ築き上げられたのです。これからは絹や桑、カイコの健康使用を目標として、活動することが重要です。

Silk is considered to be a product contrary to the SDGs for Europeans and Americans. However, mulberries, which grow fast and take also root on sloping terrain, can be a plant that produces silkworm food and also restores devastated mountainous areas.

Mulberry can be jams and juices, and the leaves can be side dishes, and teas. Until the end of the Edo period, mulberry leaves were a folk medicine for diabetes. According to recent research, it is expected that the effect moderating the rise of blood glucose level after meals will lead to the prevention of diabetes.

When grown to a large tree, the xylem becomes a wood of beautiful grain. There is a possibility that left branches after feeding and cut branches after selection can be a raw material for health foods and Japanese paper.

At the era we are living with COVID-19, various blessings of nature produced by the activities of mulberry cultivation and the sericulture performed at kindergartens, elementary schools, elderly housings, and sericulture facilities that promote mulberry cultivation and sericulture activities in depopulated areas, have the potential to create new human life.

Long history of silk could be built because it has been sustainable. From now on, it is important to work with targeting the use of silk, mulberry and silkworm for health promotion purpose.

对于欧美人来说，丝绸被认为是与可持续发展目标相反的产品。但是，生长迅速并在也丘陵地形上扎根的桑不仅可取蚕饵，可以作为修复荒凉山丘地表的植物。

果实可以制成果酱和果汁，叶子可以做配菜和茶。直到江户时代末期，桑叶一直是糖尿病的民间药。根据最近的研究，可以预期，缓解饭后血糖水平上升的影响将起到预防糖尿病的作用。当它长成一棵大树时，木质部分变成有美丽的木纹的木材。喂食后的树枝和选择后的树枝有可能是保健食品和造纸的原料。

当今电晕的时代，在幼儿园，小学，老人院及人口稀少地区的桑树种植和蚕桑生産产，这不仅幺是大自然的赐物，也可能出人类新的生活。

丝绸的悠久历史来看是因为它具有可持续性。今後，以健康使用蚕丝，桑，和蚕的目标为的很重要。

赤井 弘 / AKAI Hiromu / 赤井 弘

東京農業大学元教授 / Former Professor, Tokyo University of Agriculture / 东京农业大学前教授

国際野蚕学会 会長：日本野蚕学会 会長
President of International Society for Wild Silkmoth：President of Japanese Society for Wild Silkmoth
国际野蚕学会 主席：日本野蚕学会 主席

シルク談話会世話役代表 / Representative of Round-Table Talks of Whole Silks / 全丝绸圆桌会议代表

表紙にも記載されているように「絹大好き」のシルクは、織る、編む、着る、また食べることもできる万能の素材であり、日本では特に親しみを感じている人が多い。以前野蚕学会の座談会でシルクについて話し合ったとき、日本で作られたカイコやヤママユガ(天蚕)の生糸や布は高価でも、ぜひ着用したいとのご婦人の意見が多かった。それらを思い、日本のシルク大好きの啓発・研究支援をする日本野蚕学会の活動に加え、海外に日本のシルクの良さ及び研究活動を広く知らしめる国際野蚕学会の活動を更に深めるため、本書の3か国語表記を望みました。

一方、研究面において、カイコやシルクについて研究を始めると面白く、次々と興味が深まります。小生は学生のころ、カイコの硬い卵殻を通して外部の塩酸の刺激が内部の胚子にどのように伝わるのかに興味をもち、卵殻の構造の研究を電子顕微鏡で始めたら面白くて、知らぬ間に60年もの日々を過ごしました。

その延長で絹糸の構造や特性について研究を進め、最近、ヤママユガ科の繭糸にはフィブロイン中にライソソーム由来の多数の小孔を含み、カイコガ科の繭糸は、小孔のない緻密性繭糸であることがあきらかとなりました*。

多孔性繭糸は、布とすれば保温性が高く、高齢者の健康に良く、今後健康医療として注目されるものと思われます。

上記のカイコや野蚕のシルクたんぱく質の食べるに加え、繊維を着用して「健康に良い」となり、カイコに加え、新たに価値が見いだされてきた野蚕のシルクは、今後、シルク産業に更なる広がりをもたらす素材となるでしょう。

As stated on the cover, " Love silk" silk is a versatile material that can be woven, knitted, worn, and eaten, and many people in Japan are particularly familiar with it. When we talked about silk at a roundtable discussion at the Society for Wild Silkmoth, there were many wishes of women who wanted to try raw silk or cloth made in Japan from silkworms and saturniids (Japanese oak silkworm), even though they were expensive. With that in mind, I requested that this book would be written in three languages, in order to further deepen the activities of " International Society for Wild Silkmoth", which aims to introduce research activities of new possibilities of silk widely and internationally, in addition to the activities of "Japanesc Society for Wild Silkmoth" which supports the enlightenment and the research of Japanese silk lovers.

On the other hand, in the field of research, once I started the study of silkworms and silk, it was very interesting and my interest has been deepened one after another. When I was a student, I was interested in how the stimulus of external hydrochloric acid was transmitted to the internal embryos through the hard eggshell of the silkworm. And once I started to study the structure of the eggshell with an electron microscope, it was very interesting. After that I spent 60 years without knowing it. As an extension of this, we have been studying the structure and the properties of silkthread. Recently, it became clear that there are a large number of small holes derived from lysosomes on fibroin in case of the cocoon threads of the Saturniidae family, and those of the silkworm family are dense cocoon threads without small holes.

Porous cocoon threads have high heat retention when used as cloth, and are good for the health of the elderly, and are expected to attract attention as health care in the future. In addition to eating silk protein of domesticated silkworms and wild silkmoths mentioned above, wearing silk has been clarified to be "good for health" and, in addition, silk of wild silkmoths, which has been newly found to have value, will be a material further to expand silk industry in the future.

如封面所示，【挚爱丝绸】丝的是一种可以编织，针织，穿戴和食用的多功能材料，日本的许多人对此特别贴身。当我们在野蚕学会的一次座谈会上谈论丝绸时，有很多女性们认为希望穿戴日本制造的家蚕与天蚕（野蚕产品），尽管它们很贵。有鉴于此，为了进一步深化支持日本丝绸爱好者的启蒙和研究的日本野蚕学会的活动，以及广泛宣传日本家蚕和野蚕丝绸的新研究的国际野蚕学会的活动，我希望也这本书可以用三种语言编写：日语、英语和中文。

另一方面，在研究方面，开始对蚕和蚕丝的研究，其過兴趣越来越浓的。当我还是学生的时候，我对外部盐酸刺激如何通过家蚕的坚硬蛋壳传递到内部胚胎很感兴趣，因此开始用电子显微镜研究蛋壳的结构很有趣。在知道之前，算起来我已花了很60年的時間。我们一直在推动蚕丝的结构和特性的研究，最近，清楚地发現，Saturniidae 科的茧丝含有大量来自纤维蛋白溶酶体的小孔。另一方面，蚕蛾科的茧丝是致密性茧纱线，并没有小孔。

多孔茧丝用作布时具有很高的保温性，对老年人的健康有益，并且有望在将来作为保健衣服而培受关注。除了上述家蚕和野蚕的蚕丝蛋白可食用外，穿用丝绸也变得【有益于健康】，除家蚕之外，加之新发现具有价值的野蚕蚕丝将成为进一步地打大的丝绸行业发展并的材料。

＊赤井弘『シルク生成の謎に迫る：家蚕と野蚕はなぜ糸の構造が違うのか』化学と生物35巻12号、1997年(J-STAGE)。

＊Hiromu Akai, "Approaching the Mystery of Silk Generation: Why are Silkworms and Wild Silkmoths Different in Thread Structure" Chemistry and Biology Vol. 35, No. 12, 1997. (J-STAGE).

＊赤井弘："解决丝绸时代的奥秘：为什么家蚕和野蚕具有不同的线结构 "，《化学与生物学》1997年第35卷第12期(J-STAGE)。

長島 孝行 / NAGASHIMA Takayuki / 长岛 孝行
東京農業大学農学部教授 / Professor of Agriculture, Tokyo University of Agriculture / 东京农业大学农学部教授
デザイン農学科長 / Department of Design Agriculture Chief / 设计农业系主任
ニューシルクロードプロジェクト代表 / New Silk Road Project Representative / 新丝绸之路项目代表
日本野蚕学会副会長 / Vice-President of the Japanese Society for Wild Silkmoth / 日本野生蚕蛾丝学会副会长

地球上には10万種の繭糸動物がいて、その分だけシルクには種類があることをこれまで述べてきた。

2016年には人の手が殆どかからず、一年中完全無菌で養蚕が可能な「シルク工場」の開発に、きものブレイン㈱が成功した。中期的には100t以上の安定した最高ランクの国産無菌シルクが生産される時代の幕開けとなっている。既に機能性化粧品が商品化され、医学分野との医療素材共同基礎研究も開始されている。

一方、伝統的な桑葉を用いた養蚕は、農福連携である「訪問カイコ」養蚕が日本では開始されている。社会的弱者と呼ばれている方々が伝統的養蚕を行い、その繊維にも人気が集まっている。

桑にも新しい動きが出ている。特別な管理法、製法を開発し、美味しくて血糖値スパイクの起こりにくい「食用桑」をブランド化した。日本の代表的和菓子屋である㈱たねやもこの食用桑抹茶を原料に既に数品を販売している。

カイコ以外のシルクで特記すべきはエリナチュレ(P.59)だと思う。㈱シキボウと共同で開発したエリシルクとコットンをハイブリッドし、軽くて臭いを吸収し、紫外線遮蔽もできる「洗濯できるシルク」を開発し、2016年にはグッドデザイン賞を受賞している。この繊維を原料に産着やタオルなどの製品が販売されている。

そして2020年度は、SDGsに関連した製品開発に学生自身が取り組んでいる。以上が私達の進めているニューシルクロードプロジェクトの最近のトピックスである。

I have mentioned that there are 100,000 species of cocoon thread animals on the earth, and that there are an equal number of silk species.

In 2016, Kimono Brain Co., Ltd. succeeded in developing a "silk factory" that is completely sterile and can breed silkworms all year round with almost no human intervention. In the medium term, it is the beginning of an era when the stable production of highest-ranked domestic sterile silk can be possible in 100 tons or more. Functional cosmetics have already been commercialized, and joint basic research on medical materials with the medical field has begun.

On the other hand, for traditional silkworm sericulture using mulberry leaves, "Home sericulture", which is a collaboration between agriculture and welfare, has started in Japan. People who are called socially vulnerable people practice traditional sericulture, and their fibers are also gaining popularity.

In mulberry, a new movement has begun too. We have developed a special controlling and manufacturing method, and branded "edible mulberry" which is delicious and less likely to cause blood sugar spikes. Taneya Co., Ltd., a representative Japanese sweets shop in Japan, has already sold several products using this edible mulberry matcha as a raw material.

I think that the silk other than silkworm that should be specially noted is "Erinachure (P.59)". Under the cooperation with Shikibo Ltd., we combined Eri silk with cotton and have developed a "washable silk" that is light, absorbs odors and blocks ultraviolet rays. We won the Good Design Award in 2016. Products such as baby wear and towels are sold using this fiber as a raw material.

And in 2020, the students themselves are working on product development related to SDGs. These are the recent topics of our New Silk Road project.

我已经说过，地球上有 100,000 种茧丝动物，丝绸种类很多。

2016 年，Kimono Brain 株式会社成功开发了一家少量人工参与的完全无菌的“丝绸工厂”，该工厂可以全年饲养家蚕。从中期来看，这是一个时代的开始，在那里生产出稳定的，最高等级的 100 吨或以上的家蚕无菌蚕丝。功能性化妆品已经商业化，并且与医学领域的医学材料联合基础研究也已经开始。

另一方面，对于使用传统桑叶的蚕，日本已经开始了【家庭养蚕】，这是农业与福利之间的合作。被称为社会弱势的人，将从事传统的蚕桑业，其纤维也越来越受欢迎。

蚕桑业也有也新的动向。我们已经开发出一种特殊的管理方法和制造方法，并冠以【食用桑】brand 的商标化，这种桑的味道鲜美，不易引起血糖峰值。日本代表性的日本糖果店 Taneya Co.,Ltd. 已经出售了几种以这种食用桑抹茶为原料的产品。

我认为应该注意的是除了蚕之外的其他丝绸是 Erinachure(第59页)。通过与Shikibo Ltd.共同开发的Eri真丝和棉花杂交,我们开发了一种轻巧,可吸收异味并能阻挡紫外线的【可洗真丝】，并于2016年获得了优良设计奖。用这种原料生産的衣服和毛巾之类的产品，并已经开始销售。

2020年，是学生们已开发了与SDGs相关的产品。这些是我们新丝绸之路项目的最近主题。

中山 れいこ / NAKAYAMA Reiko / 中山 令子

図鑑作家、アトリエモレリ主宰
Author of biological picture book and president of Atelier Moreri / 生物图画书的作者和 Moreri 工作室主席

シルク談話会主宰：日本野蚕学会役員(産業情報)
Presided over of Round-Table Talks of Whole Silks
：Officer of Japanese Society for Wild Silkmoths (in charge of industrial information)
全丝绸圆桌会议主持：日本野蚕学会理事(负责丝绸行业信息)

2013年7月、赤井先生が希望されたシルクの健康使用に向けた会合「シルク談話会」を立ち上げ、赤井先生、長島先生に助けていただきながら、シルクの知識をひろめ、シルクの健康使用の入門書の執筆を始めました。

当時の私は、初めてシルクフィブロインを食べ、健康のためにシルクを着用し、製造・販売する業態の人々に取材を重ね、自分自身のシルクの知識をひろめながら、様々なシルクに出会う日々でした。

最初に編んだシルクは、藍染エリシルク平織(P.102)の裂き布。広幅(約110cm)なので2～3mあれば編めるつもりでした。1cm幅で裂いて編み始め、2mずつ買い増しをした結果、10mも使ってプルオーバーが完成(P.94左下、7分袖)。

重さ1kg強、重くて日常着にはなりませんが、時々着ています。刃物で戦う時代なら、軽い防刃着*として忍者が着たかもしれません。

一方、布を裂いたり編んだりする時に、藍の染料が飛び散ったり手についたりしたので、色止め法を知りたいと思いました。たまたま、この布で作った下着を㈱アートでキトサンSP加工をしたところ、色が固定できたので、P.103の実験につながりました。

それから様々な糸を編みました。その内、取材中に見つけたシルクの端切れを着るために、襟ぐりや袖をニットにしたトップス作ったら、身頃が伸びないので気に入っています(P.94右下、7分袖)。

この作り方なら、 今まで手を伸ばさなかった錦織などの端切れも使えると思うので、今後購入しようと思います。

In July 2013, I launched the "Round-Table Talks of Whole Silks", a meeting for the healthy use of silk that Dr. Akai requested, and with the help of Dr. Akai and Dr. Nagashima, I expanded knowledge of silk. And I started writing an introductory book on healthy use of silk.

At that time, I ate silk fibroin for the first time and wore silk for health and repeatedly interviewed people in manufacturing and in sales fields. While expanding my own knowledge on silk, I encountered various silks.

First silk I knitted is torn yarn-woven fabric of an indigo-dyed Eri silk plain weave (P.102). Since it is wide (about 110 cm), I anticipated about 2 to 3 m was enough to knit it. I started knitting by splitting it with a width of 1 cm, and as a result of additional repeated purchases by 2 m, the pullover was completed using as long as 10 m (P.94 lower left, 3/4 sleeve). It weighs a little over 1 kilogram and is very heavy, not suitable for daily wear, but it is also worn sometimes. In the era of fighting with swords, ninjas may also be able to wear them as light sword protection vests*.

On the other hand, when tearing or knitting the cloth, the indigo dye spattered or stuck on my hands, so I wanted to know how to stop the color. By chance, when the underwear made from this cloth was processed with Chitosan SP by Art Co., Ltd., the color could be fixed and this matter lead to the experiment on P.103.

Then I knit various threads. Among them, I like it because if I use the silk fragments found in the interview to make a body and knit neckline and sleeves to make a top, the body will not stretch (lower right on page 94, 3/4 sleeve length).

With this method, I think that I can use silk fabric scraps such as woven brocade that I haven't tried before, so I will buy them in the future.

2013年7月，我发起了【全丝绸圆桌会议】，这是Akai博士要求的健康使用丝绸的会议，在Akai博士和Nagashima博士的帮助下，我扩展了丝绸知识。我开始写一本介绍如何健康使用丝绸的入门书。

那时，我第一次吃丝素蛋白，为了健康而穿丝绸，采访制造和销售业务领域的人们，扩展了我自己的蚕丝知识，并見到了各种各种蚕丝。

我针织的第一条丝绸是靛蓝Eri丝绸平纹(第102页)【布条线】。因为它很宽(大约110厘米)，我预计大约2到3m就足以编织它。我开始用1厘米的宽度将进行编织，结果，由于购买了2m的物品越来越多，所以套头衫使用10m(第94页左下，3/4 袖长)完成。

它重1公斤多一点，很重，不适合日常穿着，但有时也穿。在用刀剑搏斗的时代，忍者也可以能把它们当作轻型防刀剑保护背心穿着*。

另一方面，当拆解或编织布料时，靛蓝染料会喷溅或沾在手上，因此我想要如何停止色。碰巧用Art株式会社用几丁聚糖SP处理了用这种布制成的内衣，颜色固定了，我决定在第103页上进行实验。

然后，我编织了各种线。其中，我喜欢它，因为如果我用采访中发现的丝绸碎片做一个身体，用针织领口和袖子做一个上衣，身体不会伸展(第94页右下，3/4袖长)。

用这个方法，我觉得可以用以前没试过的织锦等真丝碎片布料，所以以后会买。

＊鋼鉄に比べ弾性があるシルク、20世紀半ばまで防弾チョッキに使用。その後合成繊維に移行し、現在はクモ糸のシルクが注目されています。

*Silk, which is more elastic than steel, was used in bulletproof vests until the middle of the 20th century, after which it switched to synthetic fibers, and spider silk is now attracting attention.

*丝绸比钢更具弹性,一直被用作防弹背心,直到20世纪中叶，此后才转向合成纤维，蜘蛛丝现在引起了人们的注意。

日本には素晴らしい織物が沢山あるので、布を織るためには、技術が必要と思い尻込みするかもしれません。
　この本では、織が初めてでも簡単に取りかかれる小さな織物をテーマにしました。
　一方、1日でショールが織れる不思議な織り機で、初心者の参加が可能な桐生の織り工房(P.116)を紹介しました。私も5年くらいの間に10本近く織りました。いつでも様々な美しいシルクの糸が染上げてあるので、糸も購入しています。

　最近、知人に紹介されたシルクの編み物は、着物の裏地を裂いて作るタワシ*。シルクタワシは手に馴染んでよく汚れも落ち、自然物なのでおすすめします。考えてみれば私たちは毎日台所からマイクロプラスチックの元を流しているのです。20世紀末、琵琶湖の水の浄化のために主婦たちが使った、アクリル糸で編んだタワシ。私も2019年夏、シルクタワシに出会うまで使っていました。

　この本の執筆にあたり、多くの絹愛好家、研究者や絹製品製造メーカーの方々にご協力を頂きました。深くお礼を申しあげます。

There are many great textiles in Japan, so you might think that you need skills to weave fabrics. Theme of this book is a small loom and woven fabric that can be easily started even for the first time.
On the other hand, at Kiryu's weaving workshop (P.116), we introduced a course that beginners can participate in a mysterious weaving machine that can weave shawls in one day. I myself have woven nearly 10 shawls in about 5 years. I always sell all kinds of beautifully dyed silk threads, so I buy silk threads too.

Silk knitting that was recently introduced by an acquaintance is a tawashi* that is made by tearing the lining of a kimono. Silk tawashi is recommended because it fits in your hand well and removes dirt and is a natural product.
Come to think of it, we release source of microplastics from the kitchen every day. Tawashi knitted with acrylic thread used by housewives to purify the water of Lake Biwa around the end of the 20th century.
I also used it until I met a silk scrubbing brush in the summer of 2019.

For writing this book, I highly appreciate the cooperation of many silk lovers, researchers and silk product manufacturers cooperated.

日本有许多很棒的纺织品，所以您可能认为您需要技巧来编织织物。这本书的主题是一台小型织机和织编，即使是第一次也可以轻松启动。
另一方面，桐生的织造车间(第116页)介绍了初学者可以参加一天就能织出披肩的神秘织机的课程。我自己在大约5年内织物了近10条披肩。我总是卖各种染色精美的丝线，所以我买入丝线也。
最近熟人介绍的真丝针编织是一种tawashi*，是通过撕破和服的衬里制成的。建议使用丝绸tawashi，因为它非常适合您的手，可以去除污垢并且是一种天然产品。考虑一下点，我们每天都从厨房冲洗微塑料源。主妇用丙烯酸线编织的tawashi，在20世纪末用于净化琵琶湖的水。我也一直用到2019年夏天遇到真丝tawashi。

在书写本书時，得到了许多丝绸爱好者，研究人员和丝绸产品制造商的帮助和合作，对此我要表示深深的谢意。

シルクタワシ / Silk tawashi (scrubbing brush) / 丝绸 tawashi(擦洗刷)

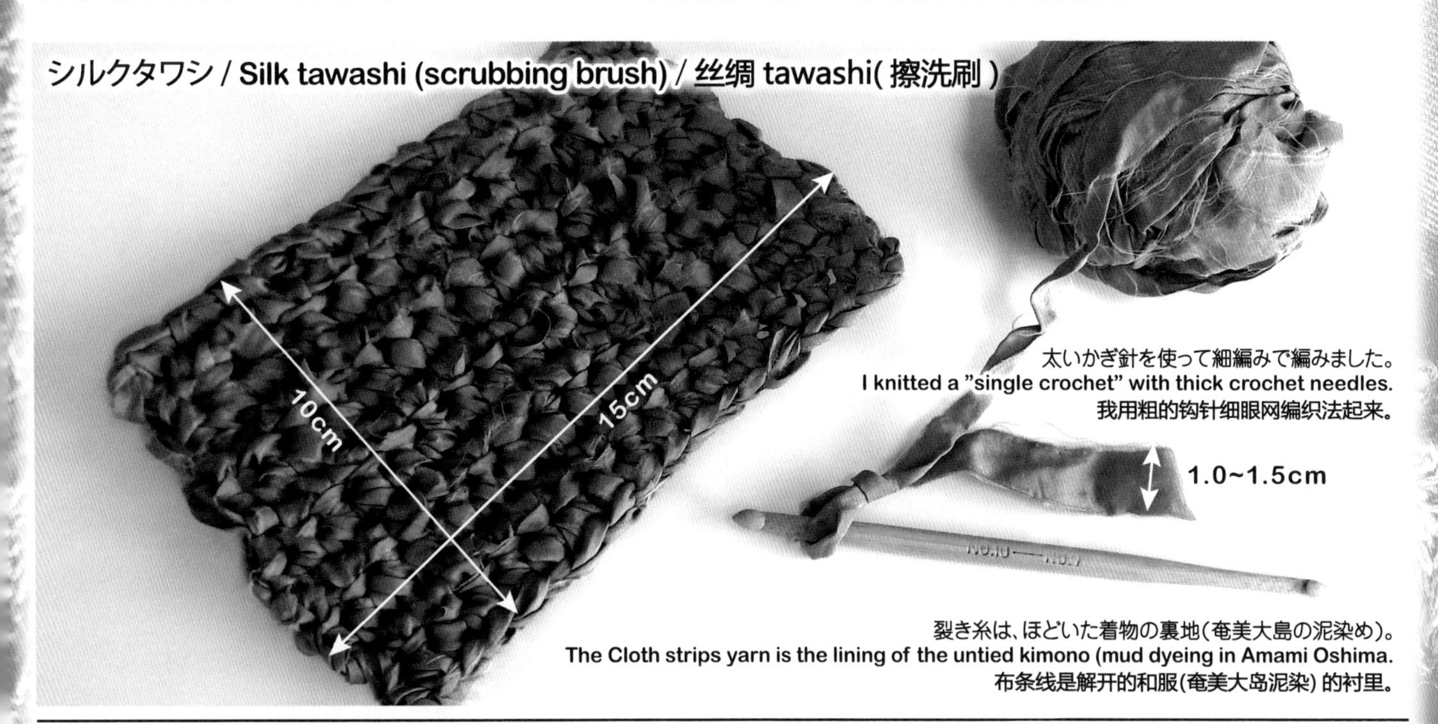

太いかぎ針を使って細編みで編みました。
I knitted a "single crochet" with thick crochet needles.
我用粗的钩针细眼网编织法起来。

裂き糸は、ほどいた着物の裏地(奄美大島の泥染め)。
The Cloth strips yarn is the lining of the untied kimono (mud dyeing in Amami Oshima.
布条线是解开的和服(奄美大岛泥染) 的衬里。

*編んでからネットで調べて、本を購入したら長編みでした。しかし、細編みがおすすめです。(2019年7月号『婦人之友』P.30～31)。

*After knitting, I bought a book by surfing and found what I knit was double crochet. However, single crochet is recommended. (July 2019 issue "Women's Friends" P.30-31).

*编织后，我在网上查了一下，买了本书，书是长编织的，但是，建议进行细眼网编织法(2019年7月发行"妇女的朋友" P.30-31)。

『絹大好き 快適・健康・きれい』
"Love silk Healthy, Comfortable, and Beautiful"
"挚爱丝绸 舒适・健康・美丽"

『絹大好き 2 まゆの秘密』
"Love silk 2 miracle of Cocoon"
"挚爱丝绸 2 茧的奇迹"

中山れいこ著作物
Works published by Reiko Nakayama / 中山 令子 出版作品

『カメちゃんおいで手の鳴るほうへ(共著)』(講談社)、『小学校低学年の食事〈1・2年生〉(共著)』(ルック)、『ドキドキワクワク生き物飼育教室』① かえるよ！アゲハ ② かえるよ！ザリガニ ③かえるよ！カエル ④かえるよ！カイコ ⑤かえるよ！メダカ ⑥かえるよ！ホタル(リブリオ出版)、『まごころの介護食「お母さんおいしいですか？」』(本の泉社)、『よくわかる生物多様性』1 未来につなごう身近ないのち 2 カタツムリ 陸の貝のふしぎにせまる 3 身近なチョウ 何を食べてどこにすんでいるのだろう(くろしお出版)、『いのちのかんさつ』1 アゲハ 2 カエル 3 メダカ 4 カイコ 5 ザリガニ 6 ホタル(少年写真新聞社)、『絹大好き 快適・健康・きれい』(本の泉社)、『虫博士の育ち方仕事の仕方(共著)』(本の泉社)、『このいろなあに はなといきもの 色覚バリアフリーえほん』(少年写真新聞社)、『都心の生物博物画と観察録Ⅰ・Ⅱ』(本の泉社)、『絹大好き 2 まゆの秘密』(本の泉社)など。

参考文献 References 参考文献

『絹大好き 快適・健康・きれい』著/中山れいこ(本の泉社)、『絹大好き 2 まゆの秘密』著/中山れいこ(本の泉社)、『いのちのかんさつ 4 カイコ』著/中山れいこ(少年写真新聞社)、『ミクロのシルクロード～目で見るシルクの生成とマユ糸の形成～』著/赤井 弘(衣笠繊維研究会)、『シルク資源シリーズ 1 黄金繭 クリキュラ』著/赤井 弘(佐藤印刷)、『シルク資源シリーズ 2 社会性の巨大繭巣』著/赤井 弘 (佐藤印刷)、『蚊が脳梗塞を治す！昆虫能力の驚異』著/長島孝行(講談社)、『一蚕とナノ技術が生んだ21世紀の新素材-メタボリックシンドロームに打ち勝つための新物質"シルクフィブロイン蛋白"』著/甲斐久美子(ヘルスビジネスマガジン社)、『シルクを食べる 絹の再発見』著/ 平林潔(高輪出版社)。『目で見る繊維の考古学 繊維遺物資料集成』著/ 布目順郎(染織と生活社)。

記載内容について参照すべき文献は、当該頁の下部 / References to the description are listed at the bottom of the page / 描述所需的文件列在页面底部

企画 Planning 规划

ブックデザイン / Book design / 书本设计：中山 れいこ / NAKAYAMA Reiko / 中山 令子
編集・構成 / Editing & Composition / 编辑与构图：アトリエ モレリ / Atelier Moreri / Moreri工作室
イラスト / An illustration / 插图：アトリエ モレリ
写真 / Photographs / 照片：ワイルドシルクミュージアム / WILD SILK MUSEUM / WILD SILK(野生丝绸)博物馆：アトリエ モレリ
P.117青梅着物博物館：井上 正行 / P.117 Ome Kimono Museum：INOUE Masayuki / P.117青梅和服博物馆：井上 正行
撮影協力 / Shooting cooperation / 拍摄合作：田中 理香子 / TANAKA Rikako
英文翻訳監修 / English translation supervision / 英文翻译监督：浦野 明人、ドクターセラム㈱ / URANO Akihito, Dr. Serum Co., Ltd.
翻訳監修中文 / Chinese translation supervision / 中文翻译监督：王 大東 / Wang Dadong
翻訳補助 / Partial translation / 翻译辅助：楊 熾源、灰原菜菜 / Gen Yang, HAIBARA Nana

絹大好き3 織る・編む・着る&食べる
Love silk 3 Weave, Knitt, Wear & Eat
挚爱丝调 3 编・织・衣与食

2021年 8月 18日 第1刷

著　者　中山 れいこ
総監修　赤井 弘
監　修　長島 孝行

発行者　新船 海三郎
発行所　株式会社 本の泉社
〒112-0005 東京都文京区水道2-10-9 板倉ビル2F
TEL：03-5810-1581 FAX：03-5810-1582
http://www.honnoizumi.co.jp
印刷／製本　音羽印刷株式会社
装　丁　中山 れいこ
制　作　アトリエ モレリ

ISBN：978-4-7807-1814-0 C2077